T0172673

Applications of Dynamic Programming to Agricultural Decision Problems

Applications of Dynamic Programming to Agricultural Decision Problems

EDITED BY

C. Robert Taylor

Routledge
Taylor & Francis Group

LONDON AND NEW YORK

First published 1993 by Westview Press, Inc.

Published 2018 by Routledge
52 Vanderbilt Avenue, New York, NY 10017
2 Park Square, Milton Park, Abingdon, Oxon OX14 4RN

Routledge is an imprint of the Taylor & Francis Group, an informa business

Copyright © 1993 Taylor & Francis

All rights reserved. No part of this book may be reprinted or reproduced or utilised in any form or by any electronic, mechanical, or other means, now known or hereafter invented, including photocopying and recording, or in any information storage or retrieval system, without permission in writing from the publishers.

Notice:
Product or corporate names may be trademarks or registered trademarks, and are used only for identification and explanation without intent to infringe.

Library of Congress Cataloging-in-Publication Data
Applications of dynamic programming to agricultural decision problems
 / edited by C. Robert Taylor.
    p. cm.
Includes bibliographical references.
ISBN 0-8133-8641-1
    1. Agriculture—Decision making—Mathematical models. 2. Dynamic programming.   I. Taylor, C. R. (Charles Robert), 1946-   .
S568.A67   1993
338.1'01'519703—dc20   93-8532
                                                                          CIP

ISBN 13: 978-0-367-01105-5 (hbk)

Contents

Tables and Figures

Tables

Figures

Preface and Acknowledgments

Perceptions about the computational "curse of dimensionality" along with a shortage of classroom treatment of applied dynamic programming have limited the application of this and other techniques for solving stochastic, multi-period optimization problems. The dimensionality curse has faded appreciably, however, due in large part to dramatic advances in computational power and in small part to refinement of the "art" of dynamic programming.

The primary purpose of this book is to demonstrate the application of stochastic dynamic programming to a wide variety of decision problems in agriculture. Articles published in this book were presented at a conference held at Auburn University on May 9-10, 1989.

The authors and editor wish to express their appreciation to Beth Rush for the outstanding desktop publishing job she did in converting all text and charts to camera-ready form. Gratitude is also expressed to Professor Earl Swanson for his additional editorial comments.

C. Robert Taylor

Contributing Authors

Timothy G. Baker, associate professor, Department of Agricultural Economics, Purdue University.

J. Richard Conner, professor, Department of Agricultural Economics, Texas A&M University.

Joan Danielson, research associate, Auburn University.

Patricia A. Duffy, associate professor, Department of Agricultural Economics and Rural Sociology and the Alabama Agricultural Experiment Station, Auburn University.

Allen M. Featherstone, assistant professor, Department of Agricultural Economics, Kansas State University.

Lee Garoian, econometrician, America Express TRS, Inc.

Russell Gum, extension specialist, Department of Agricultural Economics, University of Arizona.

Cole R. Gustafson, associate professor, Department of Agricultural Economics, North Dakota State University, Fargo.

Celia Ahrens Johnson, director of analytical services at Montana State University, Bozeman.

James W. Mjelde, associate professor, Department of Agricultural Economics, Texas A&M University.

Frank S. Novak, assistant professor, Department of Rural Economy, University of Alberta, Edmonton, Alberta, Canada.

Paul V. Preckel, associate professor, Department of Agricultural Economics, Purdue University.

Gary D. Schnitkey, assistant professor, Department of Agricultural Economics and Rural Sociology, The Ohio State University.

C. Robert Taylor, ALFA Eminent Scholar and professor of agricultural and public policy, Auburn University.

Russell Tronstad, extension specialist, Department of Agricultural Economics, University of Arizona.

Thomas P. Zacharias, director, Actuarial and Statistical Division, National Crop Insurance Services, Overland Park, Kansas.

1

Dynamic Programming and the Curses of Dimensionality

C. Robert Taylor

Stochastic and dynamic features characterize many optimization problems in agricultural decision sciences. Yet, use of numerical techniques appropriate for such problems, such as stochastic dynamic programming (DP), are not widespread. A reason often cited for the low adoption of this and other numerical techniques for solving stochastic dynamic optimization problems is what Richard Bellman, the father of dynamic programming, termed the "curse of dimensionality" (Burt). The curse of dimensionality refers to the exceedingly large computer memory and time requirements for numerically solving problems.

The theme of this article is that there is not one, but at least four curses associated with stochastic dynamic programming and other stochastic optimization techniques. These are: (a) the formulation curse, which refers to conceptualization difficulties with stochastic dynamic models of empirical problems; (b) the computational curse discussed by Bellman; (c) the curse of a complex decision rule, especially for large DP models, which means that the decision rule is difficult to visualize, rationalize, and understand; and (d) the acceptance curse which applies not only to professionals not well trained in DP, but also to potential clients (decision makers) that might use the results from such a model.

The computational curse of dimensionality, which was a central problem with DP since its development in the 1950s, has faded appreciably in the 1980s with dramatic advances in computational power and cost (Taylor). The other curses, particularly the decision rule curse, are now central problems in development and implementation of applied DP models. The purpose of this article is to discuss these curses and to suggest strategies for coping with them. Before discussing these curses, however, it may be useful for those who are not intimately familiar with dynamic programming to review this method.

Overview of Dynamic Programming

The problem of interest is to maximize the expected present value (EPV) of returns (or utility) from a T-period (stage) decision process,

$$(1) \quad \underset{X}{\text{MAX EPV}} = \underset{X}{\text{MAX}} \left[\sum_{t=1}^{T} \beta^{t-1} R(S_t, X_t, e_t) \right]$$

where $E()$ is the expectation operator, $R()$ is the one-period return function, e_t is a vector of time independent random variables, β is the discount factor, X_t is a vector of decision variables at period t, and S_t is a vector of state variables that describe the dynamics of the system. Maximization of (1) is subject to a dynamic process which can be described by the set of first-order difference (state transition) equations for the state variables,

$$(2) \quad S_{t+1} = H(S_t, X_t, \varepsilon_t)$$

where ε_t is a vector of time independent random variables. Maximization in (1) is also subject to given initial conditions, S_1.

Although DP is commonly thought of simply as a *technique* for solving dynamic optimization problems, it is more appropriate to think of it first as an *approach* to solving dynamic optimization problems, as reflected by the principle of optimality and the resulting recursive equation, and second as a *numerical* solution procedure. Dynamic programming, viewed as either an approach or as a numerical solution procedure, is based on the principle of optimality, which states:

> An optimal policy has the property that whatever the initial state and decisions are, the remaining decisions must constitute an optimal policy with regard to the state resulting from the first decision. (Bellman)

The Southern corollary to this could be stated as:

> If you don't do the best you can with what you happen to have got, you'll never do the best you might have done with what you should have had.

With an additive return function, as in (1), this principle of optimality allows one to decompose a T-period problem into T one-period optimization problems, which is reflected in the DP recursive equation,

$$V_t(S_t) = \underset{X_t}{MAX} \left[E\{R(S_t, X_t, e_t)\} + \beta E\{v_{t+1}(S_{t+1})\} \right]$$

$$(3) \qquad = \underset{X_t}{MAX} \left[E\{R(S_t, X_t, e_t)\} + \beta E\{V_{t+1}(h(S_t, X_t, \varepsilon_t))\} \right]$$

$$= \underset{X_t}{MAX} \left[W_t \right]$$

Recursively solving (3) for each value of S_t from stage T to stage 1 yields a global solution to the entire T-stage problem. Once a solution has been obtained using (3), the problem is composed, or put back together, by applying X_1^*, the optimal solution for given S_1. When S_2 has been observed (or alternatively, when the random event, ε_1, has been observed and S_2 computed with equation (2)), the optimal decisions, X_2^*, are obtained. The time paths of the state variables and the decision variables are thus traced through from the beginning to the end of the planning horizon once the random events occur. Note that in a stochastic model, time paths of the optimal decisions and the state variables cannot be traced through without values for all random variables at each stage.

The DP *approach* to solving stochastic dynamic optimization problems is reflected in the recursive equation which follows from the principle of optimality. Taking such an approach to a problem requires clear specification of state variables, decision variables, and random influences on the process. Such formulation and specification often gives insight and understanding into complex decision processes, even if the model is not explicitly solved. This is an often overlooked advantage of dynamic programming, especially compared to multiperiod programming models.

In principle, recursive equation (3) can be used to *analytically* solve the problem if all variables are continuous, the one-period and state transition equations are concave, and an interior solution exists. An analytical solution is obtained by recursively solving the first-order condition for the maximization in the recursive equation,

$$(4) \qquad \frac{\partial W_t}{\partial X_t} = \frac{\partial E\{R(S_t, X_t, e_t)\}}{\partial X_t} + \beta \partial E\{\frac{V_{t+1}}{\partial S_{t+1}}\} \cdot \frac{\partial S_{t+1}}{\partial X_t} = 0$$

beginning at the last stage and working backwards. In practice, however, mathematical functions are intractable or discontinuities exist that prevent analytical solution of all but the simplest problems.

Equation (4) may be useful for certain theoretical derivations because it *characterizes* the solution to the stochastic dynamic optimization problem. In certain cases, such a characterization can be used to derive marginal economic principles, although Ito stochastic control may give more insight into the solution than DP.

DP is typically used as a *numerical* solution procedure applied to the maximization of recursive equation (3) for discretized state and decision variables. The numerical search procedure involves searching over all values of X_t (decision alternatives) for given S_t (states). As Bellman demonstrated, application of this numerical procedure involves exponentially fewer numerical computations and comparisons that exhaustive enumeration applied to a non-DP approach using (1) and (2).

For stochastic models, it is conventional to restate (3) for discretized state and decision variables as a Markovian DP recursive equation,

$$(5) \qquad V_t(i) = \underset{k}{MAX}\left[r(i,k) + \beta\sum_j p_{ij}^k V_{t+1}(j)\right]$$

where i is a "from" index for states (defined for each combination of states of all individual state variables) and j is a "to" index for states, k is an index for decision alternatives (defined for each combination of values of all individual decision variables), and p_{ij}^k is the (transition) probability of going from state i at stage t to state j at stage (t+1) given decision alternative k. Note that the state transition probabilities are derived from (2) and the density function for ε_t.

Formulation Curse

The slowness with which DP has been adopted in the decision sciences has surprised some of the early practitioners of the technique. Burt stated that in the 1960s he thought DP would be as routinely used by now as linear programming, which is obviously not the case. He attributes the slowness with which DP has been adopted to conceptualization difficulties; that is, understanding how to formulate an empirical situation with many candidate state and decision variables as a computationally manageable DP model. Also, most economic applications of DP are sufficiently unique so as to prevent the use of canned DP solution algorithms. This has been a barrier, especially for analysts trained in mathematical programming.

As noted by Dreyfus and Law in the preface to their text and by Burt, overcoming the formulation curse requires a considerable amount of active participation by the student in formulating and solving applied problems. The implication of this for graduate training is obvious; we must offer courses on DP and other stochastic dynamics to overcome this curse. The cursory treatment given to DP in many operations research or quantitative methods courses is inadequate. For most students, a thorough understanding of DP is developed only by extensive practice formulating and numerically solving practical problems. It is encouraging that a few universities have recently instituted such training, often as a seminar series.

Motivation of students to make the large time investment required to fully understand and appreciate DP is certainly not helped by the textbook references to the curse of dimensionality associated with DP (but not mentioned for alternative techniques), and by the fact that the strengths of DP are subtle, while the weakness are obvious compared to techniques (e.g., programming approximations) commonly taught in graduate courses. One subtle strength arises when the certainty equivalence requirements are not satisfied, which means that the solution to a deterministic formulation of a stochastic problem is not exact. A second subtle strength arises because the solution to the DP model gives the *complete* decision rule, while many approximations, such as multiperiod programming models give the solution only for given initial conditions. Thus, implementing the programming formulation requires repeated solution of the model, while the DP model must be solved only once.

Curse of Dimensionality

Once the formulation curse is overcome, the computational curse may become a problem. The traditional term, "curse of dimensionality" was coined by Bellman to refer to the computational time and storage requirements necessary to numerically solve dynamic programming models with several state variables. Textbook treatments of DP as well as Bellman's books emphasize this curse so strongly that they appear to inappropriately drive away many prospective practitioners. While the computational curse clearly restricted solution of large DP models in the 1950s, dramatic advances in computational speed, storage and cost have made this curse fade rapidly in the 1970s and 1980s.

Although the curse label is usually associated only with DP, it must be recognized that the curse applies to non-DP techniques (Burt) such as use of Pontryagin's maximum principle to deterministic problems or application of Ito's stochastic control to stochastic problems. A curse exists with a control theory approach because it is necessary to analytically or numerically solve a set (one equation for each state variable) of simultaneous, often nonlinear, differential or difference equations. In the stochastic case, the set of differential equations to be simultaneously solved are stochastic, which makes the curse much more severe. In fact, Ito's stochastic control does not appear at all useful in obtaining a numerical solution (Whittle), although it may be more useful than DP for characterizing a solution.

Most textbook treatments of the DP curse of dimensionality are misleading in the sense that they usually refer to the curse in the context of the number of state variables (e.g., Bellman 1957, p. ix; Bellman and Kalaba 1965, p. 61; Hadley 1964, p. 423; Bellman 1961, pp. 94, 197), when it is actually the total number of states that limits application of the technique. The total number of states is defined as the product of the number of states for each state variable. It is the total number of states which largely determines storage requirements and computational time requirements to numerically solve a DP model.

Although the significance of the total number of states was undoubtedly clear to Bellman, wording in the literature can easily mislead beginning students of DP. The subtle distinction between the number of state variables and the number of states is quite important for two reasons. First, we often have state variables that naturally have only a few states. For example, in a flex cropping model, one state variable would be whether a field was cropped or fallowed the previous year; thus there are only two states for this state variable. Second, it has been my experience that many newcomers to DP overestimate the number of states required for most state variables in agricultural decision models. For example, we think of prices as being state variables of great significance, yet I have found that only five states for each price state variable often gives a level of accuracy in excess of the accuracy of the parameters of the model. Use of interpolation of the optimal value function in the right-hand side of (3) combined with interpolation of the resulting decision rule often gives a very high degree of accuracy.

Another aspect of the curse refers to the computer storage requirements which grow exponentially with the total number of states. Storage of all arrays used in the DP calculation (which are the one-period return function, optimal value function for all stages, optimal decision rule for all states and stages, and transition probabilities for all states, stages and decision alternatives), can easily exceed available memory on even a supercomputer for a moderately sized DP problem. However, beginning practitioners of DP need to recognize that much of this information does not have to be stored in memory. First, due to the recursive nature of the calculations, *no* elements of the optimal decision rule need to be stored (the optimal decision associated with a given state can be printed to a file when it is computed; it is no longer needed in the computations), although it is advisable to store a few elements of this matrix for efficient printout of the results. Second, we do not have to store the optimal value function for all stages; rather, we need only store the optimal value function in the previous stage in the computations (i.e., $V_{t+1}()$ on the right-hand side of recursive equation (3)), and the optimal value function for the current stage in the computations (i.e. $V_t()$), which will be used in computations for the next stage, working backwards through time.

Third, it is seldom necessary to store a fully dimensioned transition probability matrix. Most applied dynamic programming models of agricultural decisions have a mix of deterministic and stochastic state variables. A full transition probability matrix (the matrix for all states) will be comprised of large null sub-matrices which need not be stored or even looped through in computing the expected value on the right-hand side of recursive equation (3). If the stochastic state variables are independent, transition probability matrices for individual state variables can be stored to further reduce memory requirements.

Finally, if a mathematical expression is used to compute transition probabilities, storage of probabilities can be completely eliminated if memory limits are restricted. Transition probabilities can be computed as needed. Although repeated computation of the same probabilities will substantially increase computer time it will allow solution of large models on computers with limited memory, such as a personal computer.

Use of minimal computer memory required to solve a DP problem combined with the powerful and relatively inexpensive computers now allow us to solve quite large stochastic dynamic programming problems on desktop computers. With current personal computer technology it is now possible to solve stochastic dynamic programming problems with a few hundred thousand states by letting them run for a day or two.

Variants of five DP models solved on computers ranging from a 8086 personal computer to a Cray supercomputer are benchmarked in Taylor. DP models benchmarked in Taylor along with some of the very large DP models reported at this conference clearly demonstrate that present computational power, combined with efficient use of computer memory, permits us to solve quite large and realistic agricultural decision problems. Thus, the computational curse of dimensionality has faded appreciably, due largely to advances in computational power.

Large Decision Rule Curse

Unfortunately, the fading of the computational curse gives rise to the curse of an incredibly complex decision rule or decision matrix for problems characterized by several state variables. While such a decision matrix is easy to access and implement on a computer using only a sort routine (to find the optimal decision associated with the given state), the rule is difficult to visualize, digest, rationalize or understand. It is quite difficult for most of us to understand the decision rule for a problem with, say seven state variables and 200,000 states, even if there is only one decision variable. Although graphical software that is developing along with computer technology allows us to do impressive graphical presentations, we can graph in at most three dimensions which permits showing the optimal decision versus only two state variables. It is difficult to develop full comprehension of the decision rule for a problem with several state variables by only examining 2-D or 3-D slices of that decision rule. Note that the computational curse of dimensionality is determined largely by the total number of states, while the curse of the decision rule curse is largely determined by the number of state variables.

Perhaps a more troubling aspect of the curse of a large decision matrix is in knowing when the computer routines have been debugged and thus knowing when the "right" solution has been obtained. The chances of undetected bugs in computer subroutines used to compute the one-period return matrix, state transitions or transition probabilities, interpolation, and the DP search routine itself grow exponentially with the size of the problem. Just because the solution is intuitively plausible does not always mean that it is correct. Often there are several solutions that may be considered intuitive, none of which is correct. Also, some large-systems can lead to counter-intuitive solutions that are nevertheless correct.

Acceptance Curse

Dynamic programming seems to suffer from a curse of not being well accepted by many members of the agricultural community, and suffers even more of a curse in not being accepted by potential clients or actual users of a decision rule obtained by a DP model. From a professional standpoint, DP when viewed as: (a) an approach to problem solving; (b) viewed as a numerical method; or (c) viewed as a way of characterizing a solution for theoretical derivations, has not been widely adopted or accepted. Perhaps DP will be more accepted by professionals as more in-depth classroom training makes the formulation curse fade.

Acceptance rate by professionals may also increase as they recognize the normative policy content of conditional and unconditional probabilities (of different states given that the optimal decision is followed) that can be obtained by exploiting the analytical structure of the Markov process (Howard), given the optimal decision rule. There is a tendency to focus on the decision rule per se, while post-DP calculations often give information that is far more important for economic analysis.

Reluctance of potential clients (actual decision makers) to accept at face value a highly complex decision rule or matrix poses a much more fundamental problem than that of professional acceptance. Operationally, it is easy to implement the decision rule from complex dynamic programming models using only a simple sort routine on a personal computer. The user would simply input the current levels of all state variables and the computer program would return the optimal decision rule. The problem, however, is getting potential clients, very few of whom have an explicit notion of probability, to accept such a complex decision rule. This is indeed a formidable curse.

If agriculture evolves to large farms that can hire managers well trained in decision sciences, then the acceptance curse will fade. Until then, "slices" of decision rules from complex models and counter-intuitive results might serve a useful role in getting managers to "think" more carefully about their decision problem and thereby improve decisions. It is also possible to implement computer games that would pit the manager against the DP decision rule. Such a game would lead managers to more carefully think about their problem while also showing the costs of the decision rule they chose compared to the DP decision rule.

We do not have enough experience with sophisticated stochastic dynamic programming models to establish how well they might perform compared to simpler models in various situations. Although some comparisons do not show much higher expected returns with DP, a firm growth model for a representative corn/soybean farm in Illinois yielded expected after-tax returns about $20,000 per year more (about doubling annual returns) than the standard capital budgeting approach used by many firms and bankers. Such cases obviously provide a strong financial incentive for adoption that may be sufficient for potential clients to overcome the adoption curse.

Concluding Remarks

Dynamic programming remains the most promising approach and numerical solution technique for many stochastic dynamic optimization problems. The decision rule from such a model is of value to researchers and decision makers alike, but post-DP computation of probabilities of various states can be of great value for normative policy analysis. To fully exploit the potential of DP and post-DP Markovian analysis, we must provide much more in-depth training of graduate students as well as retooling of current faculty. In addition, to overcome the acceptance of results by potential clients, we must develop Extension programs that complement research efforts.

References

Bellman, M. J. *Dynamic Programming of Economic Decisions*, Springer-Verlag, New York, 1968.

_____. *Adaptive Control Processes: A Guided Tour*, Princeton University Press, 1961.

Bellman, R. and R. Kalaba. *Dynamic Programming and Modern Control Theory*, Academic Press, 1965.

Burt, O. R. "Dynamic Programming: Has Its Day Arrived?" *Western Journal of Agricultural Economics*, 7(1982):381-93.

Hadley, G. *Nonlinear and Dynamic Programming*, Addison-Wesley Publishing Company, 1964.

Howard, R. A. *Dynamic Programming and Markov Chains*, MIT Press, Cambridge, Mass., 1960.

Schnitkey, Gary D. and C. Robert Taylor. "Conventional Capital Budgeting Versus Stochastic Dynamic Analysis of Optimal Farmland Purchase and Sell Decisions." University of Illinois, Department of Agricultural Economics Staff Paper No. 87 E-390, July, 1987.

2

Representation of Preferences in Dynamic Optimization Models Under Uncertainty

Thomas P. Zacharias

Applied economists, operations researchers, and management scientists have been actively involved in the optimization of intertemporal choice problems for some time now. The applications are numerous and quite varied. In particular, agricultural applications have been shown to be quite challenging. Broadly defined, agricultural applications encompass such topics as forestry, fisheries, field crop management, farm equipment replacement, and farm-firm growth. Each of the above mentioned agricultural processes can be modeled using a systems approach. Such systems are generally characterized by intertemporal and stochastic elements. For certain agricultural systems, spatial dynamics come into play as well.

The importance of the intertemporal and stochastic nature of agricultural decision-making has been well documented (Barry; Kennedy). Moreover, casual inspection of the literature suggests that agricultural economists are keenly interested in prescriptive modeling of such systems. What has been lacking in the agricultural economics literature, however, is a discussion of the theoretical literature concerning the representation of decision maker preferences in a dynamic stochastic setting. Improper conceptualization of our intertemporal models will inevitably result in errant inference and prescription. The purpose of this paper will be to address the set of conceptual issues confronting the applied economist who is involved in prescriptive dynamic stochastic modeling at the firm-level. It is hoped that the following discussion will improve future modeling efforts and contribute to the profession's understanding of intertemporal decision making under risk -- a topic which I believe to be the essence of agriculture.

Organization of the paper will be as follows. Section II will review a set of dynamic choice literature which I will refer to as the "consistent planning" or "planning" literature. Section III will contain a review of the literature which deals with the problem of temporal resolution of uncertainty in an expected utility setting. Section IV will then discuss the implications for empirical research with particular reference to the agricultural economics literature.   Lastly, Section V will contain conclusions.

The Planning Literature

The theme of the planning literature is that of intertemporal consistency. Several suitable definitions of consistency are available. Following Hammond, one can say that "...dynamic ordering is consistent if it never leads the agent to deviate from his originally chosen plan..." (p. 166). The so-called planning literature does not explicitly treat uncertainty or incomplete information. Even so, conditions arise in which behavior might be described as intertemporally inconsistent. Inconsistency may be the result of such factors as discounting or changing tastes.

It is probably best to begin this discussion with Strotz's 1955 paper entitled "Myopia and Inconsistency in Dynamic Utility Maximization." Using the calculus of variations, Strotz defines the problem of intertemporal utility maximization and asks if a consumer will abide by or disobey his/her original consumption plan, say at t=0, if he/she can reconsider the plan at a later date.   In general, Strotz claimed the consumer will reconsider and undertake a change of plan even under conditions of perfect certainty. Strotz argued that the consumer would revise the consumption plan because of the continual updating of the discount function. Quoting Deaton and Muellbauer, "...the marginal rate of substitution between any two fixed time periods, say 1978 and 1979, will change depending upon whether they are viewed from three years before or four years before..." (p. 343). Strotz further argued that the problem could only be resolved if the discounting function was exponential. That is, a common rate of time discount would have to be applied in all periods (Mishan).

In addition to the appropriate role of discounting, Strotz addressed the issue of precommitment.   Precommitment entails excluding future alternatives in order that future behavior will conform with the desired plan of the present. At some future date one may, of course, regret his/her precommitment path.   From the standpoint of prescriptive intertemporal modeling, precommitment may be viewed as following a good heuristic.

Strotz's work seemingly lay dormant until the late 1960s when at that time a series of "planning" papers began to appear in the *Review of Economic Studies*. These studies along with Strotz's 1955 paper represent the core of the planning literature and for the most part address the fundamental sources of inconsistency in dynamic choice problems.

Pollak's 1968 paper entitled "Consistent Planning" was perhaps the initial follow-up to Strotz. In this paper, Pollak introduced the terminology naive versus sophisticated planning. Naive planning is simply a series of myopic single period optimizations which when viewed *ex post* make no sense at all. Sophisticated planning, on the other hand, essentially refers to a standard dynamic programming (DP) solution where the decision-maker allocates a given amount of stock at each stage toward his/her consumption activities so as to maximize a series of instantaneous utility functions. The sophisticated optimization is typically solved beginning with the last consumption period. Consumption in the final period is dependent upon the remaining stock with the constraint ultimately absorbed into the utility function.

In his discussion of naive and sophisticated optimization, Pollak used both a discrete and continuous time framework to reveal a flaw in Strotz's original argument as well as point out an important source of intertemporal inconsistency associated with the sophisticated optimization. For lack of a better term, Pollak focused on the "interval" problem. Two questions surface in the interval problem. These are: (1) how is optimal consumption organized within a subinterval and (2) how is the initial stock allocated among the subintervals? According to Pollak, Strotz's results implied that as subintervals converge, the conditions which determine the allocation of subinterval consumption become overriding while allocation of the stock across subintervals becomes irrelevant. Pollak provided a counter-example to Strotz revealing that allocation of the stock across subintervals was the more pressing problem as subintervals converge. Pollak did not, however, provide a general solution to the continuous decision sophisticated optimum path.

Another source of intertemporal inconsistency is changing tastes. Changes in tastes can be either endogenous or exogenous. Endogenous taste changes may be due to habit formation or choice of a particular consumption stream whereas exogenous taste changes may be due to advertising or the relationship between generations.

According to Hammond, the endogenous taste literature has focused on two basic questions. First, can valid welfare judgments be made if choices are derived from endogenous tastes? Second, does a long-run utility function exist which corresponds to long-run equilibrium choices? Hammond chooses not to address these questions, instead, he is concerned with the consistency issue in the context of endogenous taste changes and introduces the "potential addict" problem. In the potential addict problem an individual is considering the consumption of an addictive substance in a decision tree framework. Use of the substance provides initial satisfaction, however, consumption ultimately leads to addiction. Under the assumptions of the potential addict model, the individual does becomes addicted to the drug and violates an initial ordering of his/her choices via the strong axiom of revealed preference which Hammond refers to as "coherence." Hammond's "solution" to the potential addict problem is somewhat convoluted in that he resolves the problem by introducing precommitment as a new branch on the original decision tree. Hammond goes on to state that precommitment is not a mechanism for making dynamic choice consistent.

A very readable and useful discussion of the endogenous taste problem as it relates to immediate and long range planning is found in von Weizsacker.

> Another possible interpretation between short-run and long-run demand changes could be given by the theory of satisficing instead of maximizing behavior. This theory runs as follows: People set themselves certain levels of aspiration then they compare these levels with their present real situation. If this situation is inferior to the aspired level of satisfaction, they try to find ways to improve their present situation. After they have tried for a while they will either have achieved their goal or failed to do so. In the latter case, they will tend to lower their standards, i.e., the desired level of aspiration, and they will try again to attain this lower goal. If they have achieved the goal set by themselves, they become more ambitious, and they will revise upwards the goal to be achieved. It is not difficult to see that this method of satisficing - given the feedback between setting and achieving goals - will, in the long run, yield similar results to maximization.

This type of satisficing behavior would suggest a multi-attribute approach to prescriptive modeling. Discussions of multi-attribute utility applications in agricultural economics can be found in Anderson, Dillon, and Hardaker as well as Kennedy.

Blackorby et al. discuss the consistency problem in terms of separability and develop an intertemporal preference structure which may belong to a society or an individual. The relationship between consumption periods is defined by a wealth transfer mechanism (an intertemporal budget constraint) and the interaction of intergenerational preferences. Four possible societies are defined. These are: (1) the ex post society; (2) the ex ante society; (3) the altruistic society; and (4) the sophisticated society. The first three societies are naive and characterized by highly restrictive preference inheritance mechanisms or separable preferences. The consumption path of the sophisticated society is determined by a DP solution in which the choice functions of future generations are represented as a constraint in earlier periods. The authors go on to state that separability is also a problem for the sophisticated society. The work of Blackorby et al. provides an excellent conceptualization of the intertemporal choice problem, but does not provide the prescriptive modeler with any straightforward operational results.

Lastly, one apparent source of intertemporal inconsistency is the presence of durable goods. This point was made by Little who stated it as follows "...if the consumer's income and prices do not change, the condition of consistency of choice implies that exactly the same collection of things must be chosen in every period. If the period is made very short inconsistency will certainly arise, because durable consumption goods are not bought very frequently..." (p. 38-39). A discussion of this aspect of intertemporal inconsistency is beyond the scope of this paper. Suffice it to say, that the problem of durable goods along with the host of consistency problems previously mentioned should cause the applied economist to recognize the conceptual difficulties encountered when modeling dynamic choice problems.

Temporal Uncertainty and the Expected Utility Model

As stated in the introduction, agriculture is pervasively characterized by probabilistic sequential decision-making processes. The obvious question then becomes how should such a decision process be modeled if one is interested in capturing "risk aversion in an intertemporal setting?" Before proceeding further, it is probably worthwhile to review Antle's discussion of risk and dynamics in the 1983 *American Journal of Agricultural Economics*. In this discussion, Antle was able to analytically show that risk mattered in a dynamic decision model even if the objective functional had a risk-neutral structure in the Arrow-Pratt sense. An empirical demonstration of Antle's claim was provided by Zacharias et al. Antle went on to argue that improved dynamic stochastic production models using an expected value maximization criterion may be of greater value than static risk efficiency approaches that ignore the dynamics of a particular process.

A survey of the dynamic programming literature via Kennedy's text would reveal that little empirical work is available which addresses the issue of intertemporal risk aversion. The standard *ad hoc* approach has been to replace the immediate return function of a DP model with a von-Neumann Morgenstern utility function (or some variant) and then solve for the expected utility maximizing optimal policy (Karp and Pope;

Hardaker; Zacharias; Young and Van Kooten). In this section of the paper, it will be demonstrated that such a procedure is conceptually flawed for two reasons. First, application of a von-Neumann Morgenstern utility function in an intertemporal setting will result in a violation of the independence axiom. Hence, a von-Neumann Morgenstern index will not exist intertemporally. Second, risk aversion in an intertemporal setting must or should posit a preference ordering on the resolution of temporal uncertainty (Kreps and Porteus 1979).

Violation of the so-called von Neuman-Morgenstern "independence axiom" in an intertemporal setting has been known for some time by economists (Dreze and Modigliani; Markowitz; Mossin; Spence and Zeckhauser). Interestingly enough, the independence condition is not explicitly stated in the formal axioms of the *Theory of Games and Economic Behavior*. Agricultural economists have probably seen versions of the independence axiom in various sources. For discussion purposes, definitions taken from Anderson, Dillon, and Hardaker (ADH) as well as Robison and Barry are presented here. First, the ADH version:

Independence. If a_1, is preferred to a_2, and a_3 is any other risky prospect, a lottery with a_1 and a_3 as its outcomes will be preferred to a lottery with a_2 and a_3 as outcomes when $P(a_1) = P(a_2)$. Robison and Barry refer to the independence axiom as the substitution of choices axiom. Their version is as follows:

Substitution of choices: If A_1 is preferred to A_2, and A_3 is some other choice, then a risky choice $pA_1 + (1-p) A_3$ is preferred to another risky choice $pA_2 + (1-p)A_3$, where p is the probability of occurrence of A_1 or A_2.

Quoting Samuelson, the independence axiom "...says that using the same probability to combine each of two prizes with a third prize should have no "contaminating" effects upon the ordering of those two original prizes..." (p. 133). Alternatively, the independence axiom implies that the form of the preference functional is linear in probabilities (Machina 1984). That is,

(1) $U(L_c) = U[pL_1 + (1-p)L_2] = pU[L_1] + (1-p)U[L_2],$

where $U(\bullet)$ is a von Neumann-Morgenstern utility function, L_c is composite lottery of lotteries L_1 and L_2, and p is a probability, $0 \leq p \leq 1.0$. The condition in equation (1) will be violated in an intertemporal setting if the decision maker is required to make an allocation prior to the resolution of the lottery. A reasonable example of this type of intertemporal inconsistency is provided in Spence and Zeckhauser, although their numerical example is apparently incorrect.

Consider an individual with a two-commodity utility function $U = x_1^{.25} x_2^{.25}$ facing a lottery, L_1, with payoffs \$1000 and \$2000 each with probability one-half. The price vector for x_1, x_2 is (1,1). If the lottery is resolved prior to any consumption allocation, the utility function on wealth can be written $v(w) = w^5$. The certainty equivalent of this lottery is 1457.

Now suppose the individual must decide how much x_1 to purchase prior to resolution of the lottery. The individual's optimization problem can be expressed as

$$\max_{x_1} E[U] = .5 \bullet [x_i^{.25}(1000\text{-}x_1)^{.25}] + .5 \bullet [x_i^{.25}(2000\text{-}x_1)^{.25}].$$

The problem is somewhat intractable analytically, but a numerical search will yield $x_1 = 633.57$ and $E[U] = 26.23$. Denote this lottery as $L_1 = \{1000,.5; 2000, .5\}$. The certainty equivalent expression is given by

$$U(633.57^{.25} \, x_2^{.25}) = 26.23$$

Solution of this expression yields $x_2 = 746.79$ and a certainty equivalent income of approximately 1380. Denote this certainty equivalent as a degenerate lottery $L_2 = \{1380; 1.0\}$. Hence, the individual is indifferent between L_1 and L_2.

Since the individual is indifferent between L_1 and L_2, then, according to the axioms of the expected utility (EU) model, a compound lottery can be constructed with L_1 and L_2 as prizes and the individual should be indifferent between the compound lottery, $L_c = \{1380, .5; 1000, .25; 2000, .25\}$ or either of its components. In this case the individual's optimization problem is

$$\max_{x_1} E[U] = .5 \bullet [x_i^{.25}(1380\text{-}x_1)^{.25}] + .25 \bullet [x_i^{.25}(1000\text{-}x_1)^{.25}]$$
$$+ .25 \bullet [x_i^{.25}(2000\text{-}x_1)^{.25}].$$

Numeric solution of this optimization yields $x_1 = 656.57$ and a certainty equivalent income of approximately 1378 which is less than the certainty equivalent of either L_1 or L_2 - a violation of the EU model.

The preceding example demonstrates the intertemporal breakdown of the EU model. Thus, according to Kreps and Porteus (1979), "...the use of standard *atemporal* von Neumann-Morgenstern preference in any dynamic model must at least be questioned..." if decisions must be made prior to the resolution of uncertainty. The implications of the above statement for risk modeling in agriculture economics are tremendous.

In order to explicitly treat the temporal resolution problem, Kreps and Porteus (1978) have developed an axiomatic structure which they refer to

in a later publication as "temporal von Neumann-Morgenstern" (Kreps
and Porteus 1979). Their axiomatic structure is highly mathematical;
however, fairly readable numerical illustrations are provided in their
1978 *Econometrica* article. From an operational standpoint, temporal von
Neumann-Morgenstern is characterized by "two cardinal utility
functions." These functions allow for the selection of temporal lotteries
in combination with the determination of "auxiliary" decisions associated
with the dynamic choice problem.

The distinction between selection of temporal lotteries and auxiliary
decisions is discussed in Machina (1984) and not explicitly attributable to
the work of Kreps and Porteus. Auxiliary decisions are made subsequent
to the selection of a given temporal lottery. In agriculture, selection of a
particular crop cultivar could be considered a particular temporal lottery.
Fertilization or irrigation scheduling of the cultivar would then represent
the set of auxiliary decisions.

In a DP framework, selection among auxiliary decisions follows the
standard EU approach across all states. Expected utilities across states
are then ranked with the "second" cardinal index which is an increasing
function in temporal lotteries. Convexity (concavity) of the index implies
preference for earlier (later) resolution across a set of temporal lotteries.
Kreps and Porteus (1979) feel that temporal von Neumann-Morgenstern
is an "attractive" form of preference because it is simple and it allows for
"...application of standard dynamic programming methodology in
problems with this sort of criterion."

In addition to the work of Kreps and Porteus, Machina (1982; 1984) has
proposed the use of expected utility analysis without the independence
axiom. There is also the flexibility work of Jones and Ostroy along with
Epstein's work on temporal resolution of uncertainty. Discussion of these
studies is beyond the scope of this paper.

Implications for Further Research

The Planning Literature

Although the planning literature does not provide any straightforward
results which can be readily adopted by prescriptive modelers, it does
provide a basis for evaluating the conceptual validity of an empirical
dynamic choice problem. For the practicing agricultural economist,
several results seem important.

First, the interval dilemma addressed by Pollak seems to be a serious
problem. In DP terminology, the interval dilemma surfaces when one
encounters the inter/intraseasonal allocation problem. Perusal of

Kennedy's extensive literature review indicates that the profession has not adequately provided a solution procedure for the inter/intraseasonal allocation problem, even under conditions of risk neutrality. Moreover, a preference ordering leading to a cardinal utility index does not appear to be available at present to accommodate this type of problem.

Second, von Weizsacker's discussion of satisficing behavior seems to be the most operational criterion which could be readily used by prescriptive modelers. Use of target aspiration levels has a great deal of practical merit, particularly in the field of farm management. Target levels could be financial in nature (solvency or liquidity measures, etc.) and easily quantifiable. In addition, one could employ (probably without a great deal of effort) many of the numerical and analytical results which are currently available in the macroeconomic stabilization literature. As a means of comparison, one could solve both target-based dynamic optimizations and expected value optimizations. Results of the analysis could reveal potential trade-offs, trajectories of the state variables, and the variation in income.

Third, the separable nature of DP formulations should be given careful scrutiny in certain applications. The issue of functional separability has been given much attention in the duality literature. However, the implications of intertemporal separability in prescriptive models is probably not widely recognized. The implications of intertemporal separability would seem to be a major concern in the areas of natural resource and environmental economics.

Limitations of the Expected Utility Model

Applications of the expected utility (EU) model and associated risk-efficiency criteria have been widespread in the agricultural economics literature. The EU model has been and continues to be a powerful analytical tool, but it was never designed to deal with dynamic optimization problems. Thus, its use in agricultural applications must be seriously questioned on a case by case basis. If the EU model is to be used in a dynamic choice problem, then the behavioral implications should be fully explored in terms of inference and prescription. Failure to adequately address the implications of such research efforts leads to nothing more than numerical machinations.

At this point in time, temporal von Neumann-Morgentern is an operational criterion and DP applications should be pursued. Initially, these applications should probably be of low dimensionality with a great deal of attention given to sensitivity analysis. Temporal von Neumann-Morgenstern also has implications for information theory. The potential for firm-level marketing applications with an information focus would appear to be quite great.

Conclusions

This paper is basically a literature review dealing with some selected topics in the area of intertemporal decision making. The paper is not an exhaustive review nor is it intended to be. The primary motivation is to suggest that applied economists, agricultural economists in particular, carefully evaluate the behavioral underpinnings of their prescriptive intertemporal models. With regard to the stochastic DP efforts in agricultural economics, progress has been made in a wide variety of applications. However, the behavioral underpinnings have not changed appreciably since Oscar Burt's pioneering efforts in the 1960s.

It would seem as though agricultural economists have been remiss in their duty to be aware of theoretical developments in economics. The time lag between published theory and agricultural economics applications is painfully slow in some cases. The areas of applied DP and applied risk management are in desperate need of the "regenerative forces of the parent discipline..." (Harl). Agricultural economists must build stronger ties with the general economics profession. Accordingly, graduate training in agricultural economics must be strengthened.

Lastly, the conceptual issues discussed in this paper have not been fully resolved in the economics literature. The implications of these issues are relevant to literally all areas of agricultural decision-making. It is hoped that agricultural economists will find meaningful applications for existing theory and that such applications will in turn stimulate further theoretical development.

Notes

1. Strotz also raised the issue of calendar dating in the context of discounting. That is, discount weights may not change as time passes if calendar dates are important. Birthdays and anniversaries are obvious examples.

2. Although this essay is primarily devoted to a discussion of preferences which can be satisfactorily modeled for firm-level analysis, it is important to note, at least in passing, that changes in tastes can be represented by a series of succeeding generations. That is, at each instant or decision point a new generation is making decisions. At this level of generality, the conditions for consistent planning across generations become highly restrictive (Blackorby et al.).

3. Spence and Zeckhauser compute a certainty equivalent of 1376 for L_1 when a preresolution allocation must be made (p. 401-02). The authors also report x_1 = 658 for L_1 and x_1 = 688 for L_2 whereas I find x_1 = 633 for L_1 and x_1 = 690 for L_2.

References

Anderson, J.R., J.L. Dillon, and B. Hardaker. *Agricultural Decision Analysis*. Ames: Iowa State University Press, 1977.

Antle, J.M. "Incorporating Risk in Production Analysis." *American Journal of Agricultural Economics* 65(1983):1099-106.

Barry, P.J. *Risk Management in Agriculture*. Ames: Iowa State University Press, 1984.

Blackorby, C., D. Nissen, D. Primont, and R.R. Russel. "Consistent Intertemporal Decision Making." *Review of Economic Studies* 40(1973):239-48.

Deaton, A. and J. Muellbauer. *Economics and Consumer Behavior*. Cambridge: Cambridge University Press, 1980.

Dreze, J. and F. Modigliani. "Consumption Decisions under Uncertainty." *Journal of Economic Theory* 5(1972):165-77.

Epstein, L.G. "Decision Making and the Temporal Resolution of Uncertainty." *International Economic Review* 21(1980):269-283.

Hammond, P.J. "Changing Tastes and Coherent Dynamic Choice." *Review of Economic Studies* 43(1976):159-173.

Hardaker, J.B. "Optimal Management of Income Equalization Deposits." Paper presented to the Australian Agricultural Economics Society 23rd Annual Conference, Canberra, February 6-8, 1979.

Harl, N.E. "Agricultural Economics: Challenges to the Profession." *American Journal of Agricultural Economics* 65(1983):845-54.

Jones, R.A. and J.M. Ostroy. "Flexibility and Uncertainty." *Review of Economic Studies* 51(1984):13-32.

Karp, L. and A. Pope. "Range Management under Uncertainty." *American Journal of Agricultural Economics* 66(1984):437-46.

Kennedy, J.O.S. *Dynamic Programming: Applications to Agriculture and Natural Resources*. New York: Elsevier, 1986.

Kreps, D.M. and E.L. Porteus. "Temporal Resolution of Uncertainty and Dynamic Choice Theory." *Econometrica* 46(1978):185-200.

_____. "Temporal von Neumann-Morgenstern and Induced Preferences." *Journal of Economic Theory* 20(1979):81-109.

Little, I.M.D. *A Critique of Welfare Economics*. 2nd ed. London: Oxford University Press, 1957.

Machina, M.J. "Expected Utility Analysis Without the Independence Axiom." *Econometrica* 50(1982):277-323.

_____. "Temporal Risk and the Nature of Induced Preferences." *Journal of Economic Theory* 33(1984):199-231.

Markowitz, H. "Portfolio Selection: Efficient Diversification of Investments." Yale University Press, New Haven, Conn., 1959.

Mishan, E.J. "Economic Criteria for Intergenerational Comparisons." *Futures* (1977):383-403.

Mossin, J. "A Note on Uncertainty and Preferences in a Temporal Context." *American Economic Review* 59(1969):172-74.

Pollak, R.A. "Consistent Planning." *Review of Economic Studies* 35(1968):201-08.

Robison, L.J. and P.J. Barry. *The Competitive Firm's Response to Risk*. New York: MacMillan, 1987.

Samuelson, P.A. "Utility, Preference, and Probability." *Collected Scientific Papers of Paul A. Samuelson, Volume 1*, J.E. Stiglitz, Ed., Chapter 13, MIT Press, Cambridge, Mass., 1966.

Spence, M. and R. Zeckhauser. "The Effect of the Timing of Consumption Decisions and the Resolution of Lotteries on the Choice of Lotteries." *Econometrica* 40(1972):401-403.

Strotz, R.H. "Myopia and Inconsistency in Dynamic Utility Maximization." *Review of Economic Studies* 23(1955):165-80.

Weizsacker, C.C. von. "Notes on Endogenous Change of Tastes." *Journal of Economic Theory* 3(1971):345-72.

Young, D.L. and G.C. Van Kooten. "Incorporating Risk into a Dynamic Programming Application: Flexcropping." Proceedings of Southern Regional Project S-180, Savannah Georgia, March 20-23, 1988.

Zacharias, T.P. "Economically Optimal Integrated Pest Management Strategies for Control of Corn Rootworm and Soybean Cyst Nematode." Ph.D. Thesis, University of Illinois, 1984.

Zacharias, T.P., J.S. Liebman, and G.R. Noel. "Management Strategies for Controlling Soybean Cyst Nematode: An Application of Stochastic Dynamic Programming." *North Central Journal of Agricultural Economics* 8(1986):175-88.

3

Counterintuitive Decision Rules in Complex Dynamic Models: A Case Study

*James W. Mjelde, Lee Garoian,
and J. Richard Conner*

The current trend in agricultural research is to develop increasingly complex computer models. Numerous reasons ranging from increased computer skills and technology to professional acceptance explain this trend. Associated with this increasing complexity are a myriad of problems. This paper discusses one problem that may arise when using complex dynamic optimization models, that of counterintuitive decision rules.

Counterintuitive decision rules are defined as a set of decisions in which one or more of the decisions appear to be inconsistent. To clarify this definition consider the following example involving a set of decisions for four price levels. One possible counterintuitive decision rule would be: 1) buy at the lowest price level, 2) do nothing at the second lowest price level, 3) buy at the second highest price level, and 4) sell at the highest price level. As shown in this paper, such rules occasionally result from use of complex dynamic models. It is hypothesized that the complex interactions within such models lead to these counterintuitive rules.

The discussion in this paper centers around a dynamic programming (DP) model of monthly hay inventory levels for a range cattle operation. The objective of the model was to develop an effective hay inventory strategy which incorporates stochastic range conditions, production possibilities, and hay prices, along with intertemporal affects. Once developed, the model displayed counterintuitive decision rules at several stages.

To explore the potential causes of counterintuitive decision rules, this paper is organized as follows. First, background of the hay inventory decision environment is presented. Next, the hay inventory DP model is developed. Third, selected results are given. Within the results selection, several tests used to investigate potential causes of the counterintuitive decision rules are presented. Finally, questions concerning counterintuitive decision rules are posed.

Decision Environment

Range conditions vary between excellent and severe drought with weather being the primary determinant of range forage quality and quantity. The resulting uncertainty associated with forage production is an important concern of range cattle producers. This uncertainty is more dramatic when the result is insufficient forage. Strategies for combating insufficient forage include selling livestock, renting additional land, and feeding hay.

A characteristic of range forage shortages or droughts is that the duration is unknown. Selling livestock is an expensive alternative that ranchers are reluctant to use under most conditions. Availability of grazing land, associated costs, and uncertain drought duration limit the use of rented land. In this environment, feeding hay is a more appealing alternative. The response of ranchers throughout the South during the drought in 1986 vividly illustrates the reliance on hay. In drought prone areas, it is common for ranchers to carry hay inventories to meet unexpected forage demands.

Determination of optimal hay inventories concerned with uncertain forage production has been an area of study (Mauldon and Dillon; Dillon). A common approach is to use an inventory model to determine the cost minimizing level of inventory to meet a drought of unknown length. Initial inventories are acquired at the prevailing price and any shortages that occur are obtained by purchasing at a higher arbitrarily specified penalty price. Pursuing an optimal hay policy in this manner ignores all range conditions other than severe drought. In many areas of the south the probability of severe drought is much less than for excellent conditions.

Because hay prices are stochastic, the timing of hay purchases may have a large effect on costs. Also, there may be opportunities to sell inventories at a profit. Consideration of the buy and sell possibilities becomes more important when hay production is an enterprise of the ranch. In addition to buying hay, the possibility of selling and storing ranch produced hay exists.

The decision model is developed for the Texas Agricultural Experiment Station research ranch in the Rio Grande plains of south Texas. This area is characterized by a bimodal rainfall distribution with peak amounts of precipitation in June and September. Drought is defined as occurring when there are 3 or more consecutive months during the growing season in which 50% or less of the long-term monthly average precipitation is received each month. Given this definition, the area typically experiences drought in 3 out of every 10 years. The primary enterprise is a spring calving cow-calf operation on 2700 acres of rangeland. A conservative stocking rate of 27 acres per cow-unit which requires little supplemental feeding except in late winter in most years and in periods of drought conditions is assumed. Production of coastal bermuda grass hay on 20 acres constitutes a secondary enterprise. Two cuttings of hay occur during the year, June and October. Hay yield in each period is related to stochastic weather patterns. Harvested hay may be either sold at the prevailing price or stored in a 270 ton capacity facility. Hay may also be purchased and stored for future use.

Range forage shortages are met by supplemental feeding of either stored or purchased hay to meet cattle requirements. Demands for forage are calculated on a monthly basis, as a function of stocking rate, animal unit equivalents, utilization rate, organic matter intake as a percent of body weight, and number of days within the month. All parameters except stocking rate vary by month. Available forage varies by month and range condition. The difference between the demand for forage and available forage must be made-up by supplemental feeding of hay.

Range conditions are represented by a monthly index of forage production conditions based on weather (Texas Crop and Livestock Reporting Service). The index is divided into five classes ranging from excellent to extreme drought. Forage production and hay yield were related to the index values based on clipping data over several years and expert opinion (Stuth).

Specification of the Dynamic Programming Model

Successful application of DP requires the appropriate specification of stages, decision alternatives, state variables, and state variable transition equations (Bellman; Burt; Nemhauser; Kennedy). Stage length for the DP model is one month. Using months as stages provides an interval that allows for seasonal fluctuations in hay prices and forage production. The trade-off between using a monthly interval instead of a longer period is that even with presenting only the convergent decision rule, a large number of decisions must be discussed. Two stochastic state variables are included in the model, hay price and range condition. It is assumed that these two state variables are independently distributed. The inventory of hay on hand is a deterministic state variable. Decision alternatives included in the model are buying and selling hay.

Forage Requirements and Production

Within the DP model, monthly forage requirements and production are calculated. Total forage intake (FI) for month t is

(1) $FI(RC,t) = SR * W(t) * Y(SC,t)$

where RC is range conditions, SR is the stocking rate, W(t) is the cow body weight, Y(SC,t) is organic matter intake and SC is current standing crop. Stocking rate is constant at 100 cow units, but both W(t) and Y(SC,t) vary by month. Further, SC is dependent on the range condition.

Daily forage consumption per cow unit is obtained using a previously estimated organic matter intake function (Hanson). Within this function, daily consumption depends on the month, cow weight, and current standing crop. Monthly forage intake (FI) is calculated by multiplying daily forage consumption by the number of days within the month and then multiplying by the number of cow units.

The amount of range forage available (RF) is

(2) $RF(RC,t) = U(t) * SC(RC,t)$

where U(t) is the monthly utilization rate and SC(RC,t) is standing crop which depends on the range condition, RC, and month. The difference between forage intake and available range forage,

(3) $HD(RC,t) = FI(RC,t) - RF(RC,t)$

represents the demand for hay (HD) at month t as a function of range condition. In the model, a positive demand for hay must be met by supplemental feeding of either new hay production (which occurs only in June or October), current hay inventory, or purchased hay. The assumption is made that the rancher does not sell cattle in response to a drought and that supplemental feeding occurs to maintain animal performance. Meeting this demand is independent of examining decisions on optimal inventory levels. This feature of the model is part of the required actions discussed in the next subsection. A negative demand for hay simply indicates forage production is greater than forage demand and no supplemental feeding is necessary.

Recursive Equation, Inventory State Variable, and Decision Alternatives

The inventory model maximizes the present value of hay usage (production, storage, feeding, marketing) over the planning horizon. Applying Bellman's Principle of Optimization to the inventory decision process gives

(4) $\quad V_t(I,RC,P) = R_t + \max_D \{\pi_t\,(I,RC,P,D) + \beta\,E_p\,E_{RC}\,V_{t+1}(I,RC,P)\}$

where t represents the stage (month), I is the inventory level, RC is the range condition, P is the price, R_t represents net returns from required actions, max is the maximization operator, D is the decision alternative, π_t, is the immediate net returns which depend on I, RC, P and D, β is a one period discount factor, E_p and E_{RC} are conditional expectation operators taken over the price and range condition state variables, and $V_t(I,RC,P)$ is the expected value from following an optimal policy from stage t to the terminal stage. For simplicity, the stage subscripts associated with the state variables I, P, and RC have been suppressed in equation (4). The terminal value is the value of the hay inventory at the prevailing price, that is

(5) $\quad V_T(I,RC,P) = P_T * I_T$

Immediate net returns, $\pi_t(I,RC,P,D)$, includes the cost of purchasing hay and the returns from selling hay. Also included in π_t are monthly costs for producing hay. These costs are based on Texas Agricultural Extension Service Budgets. For most months the costs are zero. For months in which fertilizer is applied or harvest occurs the costs are positive. Further, harvesting costs vary by range condition. A yearly interest rate of 6% is used in calculating the monthly discount factor, β.

An unique aspect of the hay inventory model is the required actions within the recursive equation. Two components make up the required actions. The first pertains to the requirement of meeting any deficit in forage production by feeding supplemental hay. If the demand for hay is greater than new hay production plus current hay inventory levels, then hay must be purchased to meet the demand. The value of R_t in equation (4) represents the cost of purchasing this hay. The second component of R_t pertains to hay production and the 270 ton capacity of the storage facility. If the amount of hay currently in storage plus the amount of hay produced minus any hay demand is greater than 270 tons, the excess hay must be sold at the prevailing price. That is, no more than 270 tons of hay can be stored on the ranch in any given month.

Of the three state variables, only inventory is deterministic. Inventory in the next period is a function of the current inventory, required actions, current hay production and the decision alternative chosen,

(6) $I_{t+1} = f(I_t, RA_t, D_t, HP_t)$

where RA_t represents the required action and HP_t represents hay harvested in month t. Nineteen inventory levels ranging from 0 to 270 tons in 15 ton increments are considered within the model.

Twenty-two decision alternatives are included in the model. Hay purchasing can occur in units of either 15, 30 or 45 tons of hay. All purchasing of hay occurs at the prevailing price plus a $12/ton delivery and handling charge. Selling of hay occurs from 15 to 270 tons in 15 ton increments at the prevailing price. Finally, the option of neither purchasing or selling hay is allowed.

Price Transition Equation

The hay price transition probabilities were estimated using monthly state average prices for "other" (not alfalfa) hay from March, 1978 to November, 1988 (Texas Agricultural Facts; USDA). The monthly average prices were put into November, 1988 real dollars using appropriate price indices. Prices are modeled as a first-order log linear Markovian process (t-ratios in parenthesis),

(7) $\ln P_t = .30129 + .93486 \ln P_{t-1} + \alpha_t MD_t + \hat{e}_t$
           (2.23)   (30.33)
        $R^{-2} = 0.88$

where P_t is the price of other hay in month t, MD_t represents monthly dummies, α_t is the coefficient associated with month t, and $\hat{e}_t$ is the estimated error term. The t-ratio for lag price shows that a strong Markovian relationship exists. Four of the eleven monthly dummy coefficients were significant at a p-value of 20% or less. Using a procedure developed by Taylor (1981, 1984), the cumulative distribution for $\hat{e}_t$ was estimated as (t-ratios in parenthesis),

(8) $F(\hat{e}_t) = .5 + .5 \tanh(.00649 + 13.95\ \hat{e}_t).$
            $(.086)\quad (13.36)$

Combining equations (7) and (8) gives the conditional cumulative probability distribution for price,

(9) $F(P_t \mid P_{t-1}) = .5 + .5 \tanh\{.00649 + 13.95\ (\ln P_t - .30129 - .93486\ \ln P_{t-1}$
          $-\ \hat{\alpha}_t\ MD_t)\}$

Ten equally spaced price intervals (represented by their midpoint) between \$49 and \$110 per ton are modeled. This range represents the lowest and highest real hay prices observed within the data set. Finally, it is assumed the price of hay is independent of local range conditions given by the range condition indices. Regressing state average hay prices on the range condition indices showed no statistically significant relationship between prices and the range condition indices.

Range Condition Transition Equation

Estimation of the range condition transitions is complicated by the data set being incomplete. Monthly data on range condition indices for the study data were available for March, 1961 through November, 1971 and March, 1978 through July, 1987. Because of differences in weather patterns between the years within these data blocks along with potential differences in the development and collection of the indices, heteroscedasticity of the indices between blocks was suspected. Tests indicated this was indeed the case.

Further, the indices in the first block exhibited first order autocorrelation. Therefore, a slightly modified version of an estimator suggested by Theil and Goldberger was employed to estimate the range condition transition equations. This estimator is similar to the estimator suggested by Judge et al. to correct for heteroscedasticity when variances are constant within subgroups of observations.

The equation estimated is of the form,

(10)

$$\begin{bmatrix} Y_A \\ Y_B \end{bmatrix} = \begin{bmatrix} X_A \\ X_B \end{bmatrix} \alpha + \begin{bmatrix} U_A \\ U_B \end{bmatrix} \quad E \begin{bmatrix} U_A \\ U_B \end{bmatrix} = 0$$

$$E\left[\begin{bmatrix} U_A \\ U_B \end{bmatrix} \begin{bmatrix} U_A & U_B \end{bmatrix}\right] = \begin{bmatrix} V_A & 0 \\ 0 & V_B \end{bmatrix}$$

where Y is a vector of the dependent variables, X is a matrix of independent variables, α the parameters to be estimated, U the error term, V the variance/covariance matrix, and the subscripts A and B indicate the two subblocks of data. The estimator is,

$$(11) \quad \hat{\alpha} = [X_A \hat{V}_A^{-1} X_A + X_B \hat{V}_B^{-1} X_B]^{-1} [X_A \hat{V}_A^{-1} Y_A + X_B \hat{V}_B^{-1} Y_B]$$

where the $\sim$'s indicate estimates. The variance-covariance matrix of the estimator is given by,

$$(12) \quad \text{VarCov} = [X_A \hat{V}_A^{-1} X_A + X_B \hat{V}_B^{-1} X_B]^{-1}.$$

Specification of $\hat{V}_A^{-1}$ and $\hat{V}_B^{-1}$ are,

$$(13) \quad \hat{V}_a^{-1} = (1/\hat{\sigma}_A^2) \begin{bmatrix} 1 & -\hat{\rho} & 0 & 0 & 0 & 0 \\ -\hat{\rho} & 1+\hat{\rho}^2 & -\hat{\rho} & 0 & 0 & 0 \\ 0 & -\hat{\rho} & 1+\hat{\rho}^2 & -\hat{\rho} & 0 & 0 \\ \cdot & \cdot & \cdot & \cdot & \cdot & \cdot \\ \cdot & \cdot & \cdot & \cdot & \cdot & \cdot \end{bmatrix}$$

and

$$\hat{V}_B^{-1} = (1/\hat{\sigma}_B^2) \, [I]$$

where I is the identity matrix and $\hat{\rho}$ is the estimated first order autocorrelation parameter.

Application of the above estimator to the range condition indices gives (t-ratios in parenthesis),

$$(14) \quad RC_t = 16.39084 + .7402138 \, RC_{t-1} + \delta_t MD_t + \hat{e}_t$$
$$\qquad\qquad\quad (4.60) \qquad\quad (16.80)$$

where RC_t is range conditions, MD_t is a vector of monthly dummies, δ_t is the coefficient associated with month t, and $\hat{e}_t$ is the estimated error term. A strong Markovian relationship exists as given by the t-ratio associated with lagged range conditions. Three of the monthly dummies were significant at a p-value of 20% or less. Again Taylor's hyperbolic tangent procedure (Taylor 1981, 1984) was used to estimate a cumulative distribution function for $\hat{e}_t$, the corrected error terms (t-ratios in parenthesis),

$$(15) \quad G(\hat{e}_t) = .5 + .5 \tanh(-.0375 + 1.219 \, \hat{e}_t - .022 \, \hat{e}_t^3).$$
$$\qquad\qquad\qquad (-0.64) \quad\; (18.08) \quad\; (1.58)$$

Equations (14) and (15) were combined similar to equation (10) to give the conditional cumulative distribution for range conditions.

The possible range for the range conditions are between 0 and 100, with values between 20 and 96 appearing in the data set. This interval has been divided into five categories: (1) severe drought, (2) extreme drought, (3) fair, (4) good, and (5) excellent. The ranges on the indices for these five categories are 20 to 35, 35 to 50, 50 to 65, 65 to 80 and 80 to 96. Within the model, the midpoint of each interval is used to represent the range condition category.

Results

Summarizing and presenting the results of large dynamic models represents one obstacle to implementing such models as decision aids. This problem is exaggerated when dealing with a monthly versus a yearly model. With a monthly model in which the immediate returns and transition probabilities vary by month, the optimal policy will not converge between stages. The optimal policy may be year independent if the same months are compared between years. That is, the optimal policy for January in year t is compared to January's policy in year t+1, February's policy in year t is compared to February's in year t+1, etc. Thus, the optimal policy may converge between years. Even with such convergence, presenting the convergent policy for a monthly model requires twelve policies to be presented. In contrast, a yearly model requires only one convergent policy to be discussed. Development of computer decision aids which search the optimal decision table for a given set of conditions only alleviates the problem in actual use and not in presenting a model's results to an audience. Further the acceptability of a model is enhanced if results can be concisely presented.

The hay inventory model developed here is a prime example of the decision rule curse (Chapter 1). The model also suffers somewhat from the computational curse (Chapter 1). First, being a monthly model with 19 inventory states, 10 price states and 5 range condition states, the decision rule is difficult to summarize. Both state variable effects and stage effects are evident in the convergent policy. Second, the convergent decision rules suffer from counterintuitive decisions. Finally, the hay model is modest in size when compared to some DP models, but it still requires approximately four hours to solve for 10 years on an IBM AT with a math chip.

Table 3.1 is a summary of the convergent decision rules for only one of the five possible range condition states (drought range conditions), for the month of October, whereas, Table 3.2 gives the decision rules for only the fair range condition state, in December. These two tables illustrate intuitive and counterintuitive decision rules. October's policy, which is considered an intuitive policy, is summarized as follows. At the highest price hay is sold to maintain an inventory level of 30 tons, whereas, at a price of $100.85/tons hay is sold to obtain an inventory of 195 tons of hay. No hay is sold at the remaining price levels. A triangular block of buy decision is shown for low prices and all inventory levels. As price increases, hay is bought only at progressively lower inventory levels.

December's decision rule for fair range conditions represents a convergent rule which has some counterintuitive components. For example, consider a price of $82.55/ton, at the lowest two inventory levels the optimal decision is to do nothing. At inventory levels of 30, 45,

Table 3.1 Convergent Decision Rules for Drought Range Conditions for the Month of October[a]

Starting Inventory Level (tons)	Price Level ($/ton)									
	52.05	58.15	64.25	70.35	76.45	82.55	88.65	94.75	100.85	106.95
0	4	4	4	4	4	1	1	1	1	1
15	4	4	4	4	4	1	1	1	1	1
30	4	4	4	4	4	1	1	1	1	1
45	4	4	4	4	4	1	1	1	1	5
60	4	4	4	4	3	1	1	1	1	6
75	4	4	4	4	2	1	1	1	1	7
90	4	4	4	4	1	1	1	1	1	8
105	4	4	4	4	1	1	1	1	1	9
120	4	4	4	4	1	1	1	1	1	10
135	4	4	4	4	1	1	1	1	1	11
150	4	4	4	4	1	1	1	1	1	12
165	4	4	4	4	1	1	1	1	1	13
180	4	4	4	3	1	1	1	1	1	14
195	4	4	4	2	1	1	1	1	1	15
210	4	4	4	1	1	1	1	1	5	16
225	4	4	3	1	1	1	1	1	6	17
240	3	3	2	1	1	1	1	1	7	18
255	2	2	1	1	1	1	1	1	8	19
270	1	1	1	1	1	1	1	1	9	20

[a] The numbers in the table represent buy or sell decisions. Decision number 1 is do nothing. Decisions 2 through 4 are buy 15, 30, or 45 tons of hay, whereas decisions 5 through 22 represent sell decisions. The sell decisions start at 15 tons and increase to 270 tons in 15 ton increments.

and 60 tons, the optimal decision is to buy 45, 30, and 15 tons of hay. The optimal decision for the remaining inventory levels is to do nothing. Table 3.2 represents middle ground in the decision rules, where some of the other months and range conditions had more counterintuitive decision rules, while some had more intuitive decision rules.

Counterintuitive Decision Rules

The issue is "can complex dynamic models give rise to such counterintuitive decision rules or are such results strictly a function of logical errors or programming bugs?" As with most model building activities, sensitivity analysis was performed on parameters within the

Table 3.2 Convergent Decision Rules for Fair Range Conditions for the Month of December [a]

Starting Inventory Level (tons)	Price Level ($/ton)									
	52.05	58.15	64.25	70.35	76.45	82.55	88.65	94.75	100.85	106.95
0	4	4	4	4	4	1	1	1	1	1
15	4	4	4	4	4	1	1	1	1	1
30	4	4	4	4	4	4	1	1	5	5
45	4	4	4	4	3	3	1	1	6	6
60	4	4	4	4	2	2	2	1	7	7
75	4	4	4	4	1	1	1	1	8	8
90	4	4	4	4	1	1	1	1	1	9
105	4	4	4	3	1	1	1	1	1	10
120	4	4	4	2	1	1	1	1	1	11
135	4	4	4	1	1	1	1	1	1	12
150	4	4	3	1	1	1	1	1	1	13
165	4	4	2	1	1	1	1	1	1	14
180	4	4	1	1	1	1	1	1	1	15
195	4	3	1	1	1	1	1	1	1	16
210	4	2	1	1	1	1	1	1	1	17
225	4	1	1	1	1	1	1	1	1	18
240	3	1	1	1	1	1	1	1	5	19
255	2	1	1	1	1	1	1	1	6	20
270	1	1	1	1	1	1	1	1	7	21

[a] The numbers in the table represent buy or sell decisions. Decision number 1 is do nothing. Decisions 2 through 4 are buy 15, 30, or 45 tons of hay, whereas decisions 5 through 22 represent sell decisions. The sell decisions start at 15 tons and increase to 270 tons in 15 ton increments.

model to check for logical or programming errors. For example, the cost of hay delivery was varied from a negative value to a positive value. Changes in the decision rules resulting from varying this parameter were intuitively correct. That is, the optimal decision rules reflected a decrease in the cost of buying hay (a negative delivery charge) by buying relatively more hay. An increase in the delivery charge resulted in less hay being purchased. Such sensitivity analysis and debugging procedures do not guarantee an error free model, but no errors were found. The issue remains can a complex dynamic model produce counterintuitive decisions?

A closer study of the inventory model is revealing in examining the hypothesis that complex models can give rise to counterintuitive decision rules. Within the model, many different components are interacting at each stage. Potentially, it is these interactions which give rise to the counterintuitive decisions. First, both the price and range condition equations were estimated using monthly dummies. In 7 of the 12 months, the signs of the coefficients associated with the monthly dummies in each equation are identical. Thus, conflicting signals are passed to the decision model. For example, consider a positive monthly dummy sign for next month in both equations. For the price equation the positive sign indicates that the probability of higher prices next month is increasing, suggesting that purchasing of hay in the current month may be profitable. But a positive sign associated with the range condition equation indicates a higher probability of improved range conditions, suggesting that purchasing of hay is not necessary. To test the hypothesis that the monthly dummies were causing the counterintuitive decisions, two modified models were developed. In model one, all monthly dummies associated with the price equation were set equal to zero, whereas, in model two, all monthly dummies associated with both the price and range condition equation were set equal to zero. Resultant convergent decision rules for both models had much fewer counterintuitive components than the full monthly affect model, but some counterintuitive components still remained.

A second aspect of the model which was hypothesized as associated with the counterintuitive decisions was hay production costs, demand for hay, and yield of harvested hay varying by month and by range condition within a month. Of these three factors, only the demand for hay was tested. The inventory model was run with zero demand for hay in all range conditions and months. The resultant decision rules were as expected. For all prices except the lowest price, all hay was sold. This indicates several factors: (1) buying hay and storing it for resale is not a profitable enterprise, (2) when at the lowest price level, the probability of receiving a higher price next month is quite high, and (3) the varying demand for hay by month and range condition may be a contributing cause to the counterintuitive decisions. From these results, it can be inferred that including varying costs and hay yield contribute to the counterintuitive decisions.

Finally, linear interpolation which is used on the deterministic inventory state variable because both demand for hay and hay yield are not in units of fifteen tons, is hypothesized as a contributing cause of the counterintuitive decisions. Following Burt's (1982 p. 387) reasoning that "usually linear interpolation is adequate if the intervals are not too wide," the inventory interval width was decreased from 15 to 10 tons. This change almost doubled the computational time and resulted in only a few less counterintuitive decisions. To further test if linear interpolation was a contributing factor, the model was solved without linear interpolation. A change in the convergent decision rule was noticed in that more hay was sold and less bought, but counterintuitive decisions were still prevalent. An observation from the senior author's homework assignment for a DP class supports the hypothesis, linear interpolation may cause counterintuitive decision rules. The decision rule for a simple yearly DP model with deterministic nitrogen carryover using linear interpolation and stochastic price flipflops between stages. That is, in one state the optimal decision is to apply 76 lbs. of nitrogen in stage t, 74 lbs. in stage t+1, 76 lbs. in stage t+2, 74 lbs. in stage t+3, etc. Personal communication with Robert Taylor (1989) indicates that he has experienced the same flipflopping of decisions when using linear interpolation. Some bias is present when using linear interpolation and this bias may cause this flipflopping.

All the sensitivity analyses, debugging, and examining of various model formulations does not prove complex dynamic models may lead to counterintuitive decisions, but it does support this notion. Two sets of monthly dummies affecting independent stochastic transition equations, varying production costs, hay yield, and demand for hay, discretizing continuous state variables, and linear interpolation most likely interact to give counterintuitive decisions. The question becomes, "are complex dynamic models useful if such counterintuitive results are possible or are such models just an exercise in futility?"

Concluding Remarks

The question remains "was developing the hay inventory model an exercise in futility or was knowledge gained from the exercise?" We definitely feel knowledge was gained. Most importantly, the model reveals that complex dynamic models may lead to counterintuitive results. As computer technology increases the tendency has been to increase the complexity of decision models to fit the computer capacity. This trend will continue both to gain knowledge about complex systems and for professional acceptance. The trend has been away from what Burt (1982 p. 383) states as the primary objective in modeling "to capture the essential aspects of the phenomenon under study and yet keep the model as simple as possible." Further, the counterintuitive aspects should not be used as an excuse for developing a static or very simplistic dynamic model. This aspect is a potential reality, but appears to be more so in monthly or quarterly models than in annual models. In most annual models, immediate returns and transition probabilities are identical between stages in a given state. This is usually not the case in a monthly or quarterly model. Care must be used in developing complex models. For example, the results indicate that techniques, such as used by Mjelde, Taylor, and Cramer, to smooth transitions between months instead of monthly dummies should be considered.

Another issue is, "was any knowledge gained about the original problem, optimal hay inventory levels?" Again, the answer is yes. Several generalities from the convergent decision rules are obtained. Ignoring the counterintuitive results, the decision is to buy hay only at low prices and/or low inventory levels. In general, the model indicates hay should only be bought, except as required for supplemental feeding, at or below a price of $76.45/ton. Except in the two production months, June and October, hay should be only sold at a price of $100.85/ton or higher. In the two production months, hay is sold at lower prices but only at high inventory levels. Besides price, the decision rules show both range conditions and monthly effects. Further, it is not profitable to buy hay and store it for resale at a later date. This is intuitively pleasing because hay management is for minimizing the cost of supplemental feeding and is not a buy/sell enterprise. This model was intended as a first step in building a hay inventory model, plans to refine the model using a forage simulation model are currently under development. It appears from examining the decision rules that an inventory model which included only the production months and possibly one additional month in which hay price is usually higher as stages would be adequate. Such a simpler model should be adequate for management of hay inventories.

Finally, the above discussion raises several points to ponder, all necessitating additional research. First, is there any value to models which give counterintuitive decision rules? The answer to this question is that the value of such models will depend in part on the relative amount and causes of the counterintuitive decisions. Second, should a researcher strive for strictly intuitive results? The answer is yes within bounds. A researcher should not sacrifice model integrity, theoretical considerations, etc., to eliminate counterintuitive decisions. Finally, will models with counterintuitive decisions be accepted by decision makers, peers, and professional journals? The answer to this depends in part on the answers to the first two questions. A decision maker will be interested in the costs/benefits associated with following a counterintuitive policy versus his current decision policy. Professionally because of pressures to publish, lack of acceptance may force researchers into selective reporting of results. Counterintuitive decision rules appear to be a reality which needs to be dealt with openly.

References

Bellman, R.E. *Dynamic Programming.* Princeton: Princeton University Press, 1975.

Burt, O.R. "Dynamic Programming: Has Its Day Arrived?" *Western Journal of Agricultural Economics* 7(1982):381-94.

Dillon, J.R. "Inventory Analysis of Drought Reserves for Queensland Graziers: Some Empirical Analytics." *Australian Journal of Agricultural Economics* 6(1962):50-67.

Hanson, D.M. "Influence of Contrasting Prosopis/Acacia Communities on Diet Selection and Nutrient Intake of Steers." Unpublished M.S. Thesis, Texas A&M Univ., 1987.

Judge, G.G., W.E. Griffiths, R.C. Hill, and T. Lee. *The Theory and Practice of Econometrics.* New York: John Wiley and Sons. 1980.

Kennedy, J.O.S. "Application of Dynamic Programming to Agriculture, Forestry, and Fisheries: Review and Prognosis." *Review of Marketing and Agricultural Economics* 49(1981):141-73.

Mauldon, R.G. and J.L. Dillon. "Droughts, Fodder Reserves and Stocking Rates." *Australian Journal of Agricultural Economics* 4(1959):45-57.

Mjelde, J.W., C.R. Taylor and G.L. Cramer. "Optimal Marketing Strategies for Wheat and Corn Producers in the 1982 Farm Program." *North Central Journal of Agricultural Economics* 7(1985):51-60.

Nemhauser, G.L. *Introduction to Dynamic Programming.* New York: John Wiley and Sons, 1966.

Stuth, J.W. professor, Department of Range Science, Texas A&M University, College Station, TX. Personal Communication. 1988.

Taylor, C.R. professor, Department of Agricultural Economics, Auburn University, Auburn, Alabama. Personal Communication. 1989.

_____. "A Simple Method for Estimating Empirical Probability Density Functions." Dept. Agr. Econ. Pap. 81-1, Montana State University, 1981.

_____. "A Flexible Method for Empirically Estimating Probability Functions." *Western Journal of Agricultural Economics* 9(1984):66-76.

Texas Agricultural Extension Service Budgets, Texas A&M University, College Station, TX. 1988.

Texas Agricultural Facts. Texas Agricultural Statistical Service, Austin, TX. Appropriate Issues.

Texas Crop & Livestock Reporting Service. Texas Agricultural Statistical Service, Austin, TX. Appropriate Issues.

Thiel, H. and A.S. Goldberger. "On Pure and Mixed Statistical Estimation in Economics." *International Economics Review* 2(1961):65-78.

USDA, SRS. Agricultural Prices: Annual Summary, Appropriate Issues.

4

Optimal Stochastic Replacement of Farm Machinery

Cole R. Gustafson

A problem of considerable interest in agriculture economics involves optimal replacement of depreciable farm assets. Numerous studies have investigated the effects of tax policy, inflation, and remaining salvage values on optimal replacement periods by calculating the present value of costs over an infinite horizon for each possible replacement year and then selecting the year with minimum cost (Bates, Rayner and Custance; Bradford and Reid; Chisholm; Kay and Rister; Leatham and Baker; Perrin; Reid and Bradford).

While these studies have contributed greatly to our knowledge of replacement behavior, the solution method employed is limited in three respects. First, a uniform replacement strategy whereby every asset in the sequence is replaced at equal intervals may be sub-optimal. Even though replacement problems are usually stationary (with respect to states, actions and returns), initial and ending conditions generally warrant a departure from repetitive policies. In such instances, dynamic programming is the preferred algorithm of search because each replacement is permitted a unique age.

Second, these studies do not consider the stochastic environment in which replacement decisions are made. Any variable affecting the replacement decision, including net returns, machine performance, asset prices, and capital costs, may be regarded as uncertain. Weersink and Stauber, and Burt (1965) utilize stochastic dynamic programming to consider both uncertain revenue and the possibility of machine failure. In the event of failure, Burt assumes replacement with a new asset whereas Weersink and Stauber assume overhaul of the existing machine. Finally, previous studies assume unlimited capital availability. The inclusion of a capital constraint adds realism to the analysis and provides additional insight into investment behavior under uncertainty. Capital may be rationed either internally or externally. Internal rationing may permit more timely investment if fund availability varies, allow the firm to take advantage of unforeseen investment opportunities or reduce costs when unexpected failure occurs.

This paper formulates a cashflow-constrained stochastic dynamic programming model of optimal farm machinery replacement. The model's solution illustrates the tradeoffs between the conflicting objectives of minimizing expected present cost of machinery services over a finite planning horizon, limiting unanticipated machine downtime, and maintaining cash balances. Sources of uncertainty in the decision environment include variable investment fund availability and untimely machine failure. Following sections develop the model and apply it to a combine replacement decision on a western Minnesota cash grain farm.

Replacement Model

Optimal replacement of farm machinery is critical to the financial management of modern, capital-intensive agricultural firms. The decision involves selection of a series of assets which minimize expected present costs of providing a constant flow of services over a finite planning horizon, N.[1] At each stage or period, the firm must decide whether to replace an existing asset. If replacement is the long-run least cost alternative and sufficient cash reserves are available, a new machine is purchased. Otherwise the existing machine is retained for another period.

In addition to planned replacement, involuntary replacement of farm machinery occurs. When a machine unexpectedly fails, the firm can either (1) repair the existing machine and incur its operating costs and chances of failure in future periods, or (2) involuntarily trade the failed machine for a new one with anticipation of lower operating costs and chances of failure at some future date. Hence, simultaneity exists because involuntary decisions are contingent on planned replacement decisions, and vice versa.

At any stage k of the horizon, the firm's decision is a direct function of the system's state as reflected by machine age j at the start of period (e.g. age is assumed to be a proxy for productivity and condition). Primary factors influenced by age include:

C_j = annual after-tax operating and ownership costs of a machine which is age j, including fuel, repairs, financial payments, insurance, tax depreciation deductions, and interest on equity capital;

$PROB_j$ = the conditional probability that a machine of age j, in working order, fails during the period;

$FAIL_j$ = the minimum expected cost of repairing or replacing a failed machine of age j, subject to cash availability; and

$TRADE_j$ = net costs of trading the incumbent machine, in working order, for a new machine.

At the first stage of the horizon, the system's state is evaluated and a decision is made whether or not the incumbent machine should be replaced. In each subsequent period, the problem reoccurs and the decision is made again. Each time, the decision is based solely on information available at that time (i.e. it does not depend on replacement decisions in previous periods). This recursive aspect of the problem lends itself to the solution algorithm of dynamic programming.

Following Dreyfus and Law, the above replacement problem is specified in terms of a discrete stochastic dynamic programming model:

$$(1) \quad S_{j,k} = \min \begin{Bmatrix} KEEP = C_j + PROB_j FAIL_j + (1 - PROB_j)\beta S_{j+1,k+1} \\ BUY = TRADE_j + C_0 + PROB_0 FAIL_0 + (1 - PROB_0)\beta S_{0,k+1} \end{Bmatrix}$$

s.t. Cashflow > 0

where $S_{j,k}$ equals the minimum expected cost of the remaining process if we begin year k with a working machine of age j. The real after tax discount rate β equals $1/[1+(1-t)i]$, where i is the real opportunity cost of equity capital and t is the marginal tax rate.

The model is evaluated over all remaining stages of the horizon $k=1,...,N-1$ and possible asset ages, $j=1,...,k-1$, including the special case that occurs if the original asset is never replaced, $j=y+k-1$, where y is age of that asset at the beginning of the analysis. This notation and previous variable definitions imply replacement decisions are made at the beginning of each stage and that the process terminates at the beginning of the last period of the planning horizon.

Because the system reverts to state $j=0$ every time replacement occurs, the model is regenerative and considered as a special class of dynamic programming models (Dreyfus and Law). The age of the asset increases by one and the system transits deterministically from stage to stage unless a new machine is purchased, in which case the status of the system regenerates.

Technology and all other functions comprising the model are assumed stationary with respect to time. Hence, replacement always occurs with machines of similar type.

Planned Replacement

At the beginning of stage k, the firm evaluates the expected cost of retaining, *KEEP*, and replacing, *BUY*, the incumbent asset. If retained, total expected costs over the remaining horizon equal immediate costs of operating and owning the incumbent asset in the present period, plus the expected costs of either fixing or replacing the machine if it fails, and the minimum expected present cost of the remaining process, which assumes the next stage ($k+1$) begins with the current asset, albeit one year older ($j+1$). The remaining process may or may not involve replacement of the existing machine.

If a new machine is bought, the following net costs of trading are incurred:

(2) $TRADE_j = DOWN-R_j+DEPR_j+RBAL_j+CAPGAIN_j$

where *DOWN* is the downpayment required to purchase the new machine,[2] R_j is the remaining value of a working machine, $DEPR_j$ is amount of depreciation recapture due, $RBAL_j$ is the remaining loan balance on the current machine, and $CAPGAIN_j$ is the capital gains tax due. Thus, if the existing machine is replaced, total expected costs over the remaining horizon equal net costs of trading, plus immediate operating and ownership costs of a new machine, the expected costs of either fixing or replacing a new asset that fails in period one, and the minimum expected present cost of the remaining process, which assumes the next stage ($k+1$) begins with an asset of age 1.

After determining present costs of keeping the existing machine and buying a replacement, the minimum cost alternative would normally be selected. However, cashflow before machine costs in agriculture is highly variable, and a likelihood q exists that sufficient funds will not be available to purchase a new machine.[3,4] In these instances, the firm must defer replacement to a subsequent year and keep the existing machine. As a result, the decision rule for minimizing expected present costs of keeping and buying a new replacement becomes:

(3) $S_{j,k} = min$ $[KEEP,BUY] = min[KEEP, (\{1-q\}BUY+qKEEP)]$

s.t. Cashflow > 0

where q reflects the firm's subjective expectation of a binary state variable $\{0,1\}$ describing the unavailability of funds for machinery replacement at each stage of the process.

This objective function does not consider the risk consequences of possible decision rules. Such a formulation would involve a whole farm analysis and abstract from the focus of this study (Burt, 1988). In addition, the distribution of costs resulting from various individual replacement decisions is not expected to substantially alter the risk position of the entire firm.

Involuntary Replacement

Services provided by farm machinery are susceptible to disruption when fire, flood, accident or failure of a major component occurs. Some of these events are insurable, while others are not. Those that are not constitute a risk for the firm. When an uninsured catastrophic event occurs, the firm must decide whether to repair the existing machine or purchase a replacement. However, in addition to direct costs of restoring the machine, firms incur a loss associated with downtime and the disruption of services. To mitigate downtime losses, a custom operator may be hired to complete the operation or a temporary machine may be obtained on a rent- or lease-basis.

Continuing the development of the replacement model above, assume:

$$(4) \quad FAIL_j = \min \begin{Bmatrix} REPAIR = D_j + FIX_j + \beta S_{j+1,k+1} \\ REPLACE = TRADEF_j + \beta S_{0,k+1} \end{Bmatrix}$$

s.t. Cashflow > 0

where $FAIL_j$ is the minimum present expected cost, subject to fund availability, of an optimal policy of repairing or replacing a failed machine which was age j at the beginning of the period; D_j are the downtime costs associated with the failure; FIX_j is the unexpected cost of repairing the failed asset and $TRADEF_j$ is cost of exchanging the failed asset for a new machine defined as:

$$(5) \quad TRADEF_j = DOWN\text{-}T_j + DEPR_j + RBAL_j + CAPGAIN_j.$$

$TRADEF_j$ is greater than $TRADE_j$ due to the lower remaining value of an unrepaired machine, T_j. The probability of machine failure, $PROB_j$, was defined above to be conditional on age. The probability that a machine does not fail until age m is described by repeated application of the multiplicative law of probability:

$$(6) \quad \text{PROB} = \prod_{j=1}^{m} \pi$$

In a sense, the firm's decision to involuntarily replace a machine is conditional on similar factors influencing planned investment, including relative future operating costs, remaining values, capital costs, and cash availability. Repair and retention of the existing asset implies the existing machine with its pattern of repairs will be used in following periods. However, unique factors affect involuntary replacement as well. If the failed machine is repaired, downtime costs are incurred in addition to costs of fixing the machine. If the failed machine is traded unrepaired, it is assumed the new machine is available immediately and downtime costs are negated. Hence, the decision to involuntarily replace is contingent upon total costs of fixing, level of downtime incurred, and value received for the failed machine.

Solution Procedure

The procedure for solving the replacement problem begins with specification of boundary conditions for the last period. Boundary conditions for this problem are determined by replacing S_{jk+1} in the right hand side of equation (1) with the remaining value of the asset realized at the end of the planning horizon, R_j if the asset is in working condition and T_j if the asset failed.

Working backward stage by stage from the end of the horizon, the remainder of the problem is recursively solved according to the algorithm of dynamic programming and principle of optimality. At each stage, expected present costs of keeping the existing machine and buying a replacement are calculated for every possible state of the system. An important part of the decision involves future benefits of the present decision, S_{jk+n} (the minimum cost of providing machine services from period $k+n$ to the horizon). But, because the optimal decision is derived by working backward, the value of S_{jk+n} is already known. Given this information and expected costs of the current period, the firm can choose the minimum cost option of either keeping or replacing the current machine and assign the respective value to S_{jk}, since the principle of optimality assures that no nonoptimal continuation from year $k+1$ to N need be considered (Dreyfus and Law). The decision for remaining stages is independent of historical replacement policies; it depends solely on current and future information. Eventually the firm arrives at $S_{j,1}$. The resulting sequence minimizes the expected present cost of machine services until the horizon.

Application

The above model was applied to a combine replacement decision on a typical cash grain farm in western Minnesota. A panel comprised of six machinery dealers and three farmers in the vicinity of Ottertail County provided necessary empirical data to estimate the replacement model.[5] Medium-sized combines in the area are used an average of 300 hours per year for harvesting corn, soybeans and small grains. The purchase price of such a combine was $80,000 in the spring of 1989.

Operating Costs

Variable costs of operating a combine include fuel, lubrication, labor, repair, and downtime losses that occur when the machine unexpectedly fails. Fuel, lubrication and labor costs were calculated by formulas provided in the Agricultural Engineers Yearbook. Total costs of repairing a combine consist of routine repair and maintenance costs as well as costs associated with a major breakdown. With the exception of costs incurred during the first year of operation, which are presumed to be covered by warranty, total costs of repairing the combine in this study were based upon recent survey data (Hardesty and Carman) and allocated between the two categories in the following manner.

First, service managers at each of the panel dealers in this study were informally asked to describe the nature of untimely breakdowns. The average cost of such a breakdown was reported to be $7,000. Next, they were asked to subjectively estimate the probability that such a breakdown would occur each year of the combine's useful life. These probabilities were estimated to be 5 percent in year one of operation and up to 50 percent by year ten. Interestingly, the estimated probabilities were not conditional on the timing of previous breakdowns. Service managers noted that modern combines consist of numerous independent systems (i.e. engine, hydraulics, final drives, thrashing components) and that a failure of one does not influence the probability of another system failing. Further, replacement parts are also subject to failure.

Expected costs of a major breakdown were derived based on the above responses. These costs were then subtracted from total repair costs to yield routine costs of repairing and maintaining the combine. To assure accuracy, the panel of participating dealers and farmers were asked to review their responses. Each stated the pattern of routine costs appeared reasonable when the value of the farmer's labor and investment in tools and repair facilities were included.

The final component of operating cost reflects downtime losses when a combine unexpectedly fails. Downtime hours for this study are specified as a function of cumulative machine hours (Hardesty and Carman). In the event of failure, dealers and farmers on the panel reported that a replacement machine is usually leased (95 percent frequency) rather than hiring a custom operator to complete the harvest. Typical lease rates are $75 per hour of machine use. Only in exceptional cases are loaner combines provided by dealers.

Ownership Costs

Costs of owning a combine include depreciation, payments of principal and interest, an opportunity cost on equity capital, insurance, and taxes. The remaining value of the combine at the end of each period is an important determinant of these costs. Bradford and Reid note the limitations of previously-derived empirical functions. Therefore, the following remaining value function is postulated based upon responses of the panel:

$$R_j = .85(.90)^j$$

If the owner decides to trade rather than repair a failed machine, the remaining value of the failed machine T_j is assumed to be 95 percent of the asset's net value:

$$T_j = (R_j\text{-repair costs})\text{x}.95$$

which reflects the owner's weakened bargaining position. However, trading a failed machine mitigates downtime costs.

If the asset is financed with debt capital, annual cash payments of interest and principal are incurred. Typical arrangements include a downpayment of at least 33% and a repayment period of five years. On remaining equity capital, an opportunity cost is assigned. Real before-tax interest rates on equity and debt capital are assumed to be 4 and 6 percent, respectively.

The annual premium for insurance on the combine is assumed to be $.004 multiplied by the remaining value of the combine. In the event of loss, the indemnity is assumed to cover all losses, including downtime.

For tax purposes, the asset is depreciated over seven years according to the double-declining method permitted under 1989 Modified ACRS provisions. It is assumed the owner deducts the allowable Section 179 limit of $10,000 in the first year of use. Under 1989 provisions, investment tax credit is not available. Excess depreciation claimed is recaptured at the time of replacement. Based upon the financial characteristics of farmers in the area, a marginal tax rate of 15 percent is assumed (Spengler). All operating costs and interest on borrowed funds are deductible for tax purposes. The length of planning horizon is assumed to be 30 years.

Results

The expected patterns and present costs of optimal combine replacement strategies with and without financing are shown in Table 4.1. In general, the first combine purchased is retained longer than succeeding replacements. Without debt financing, the initial combine is retained 10 years and then replaced every 4 years over the remaining horizon. The expected cost of such a policy is $239,498 over the 30 year horizon.

When the combine is financed with 67 percent debt capital, the cost of the optimal policy is $69,599 greater than exclusive equity financing because the real cost of borrowing is assumed to be 2 percent greater than the discount rate. With debt financing, the ownership period for both the first and succeeding machines increases by two years.

Comparing optimal replacement strategies with farmers' actual practices is one means of model validation (Reid and Bradford). Previous

Table 4.1. **Expected Pattern and Present Cost of Optimal Combine Replacement Strategy**

	Level of Financing	
	0%	67%
Combine purchased in years:	0	0
	10	12
	14	18
	18	24
	22	
	26	
Expected present cost	$239,498	$309,097

strategies yielded by non-dynamic programming algorithms appeared longer than ages actually used by U.S. farmer tractor owners. Given a policy range of 4 to 12 years in this study, it is important to compare systems with similar states, stages, and ending horizons. The sensitivity of an optimal policy can be easily examined by comparing the value of *BUY* and *KEEP* at each stage. In the case of complete equity financing, the additional present cost of trading the existing asset one year prior to the optimum (i.e. replace in year 9 rather than 10) is only $1,851. However, this differential increases markedly to the point where a nonoptimal trade after only one year of use costs $11,953.

The present cost and sequence of purchases associated with an optimal replacement policy are conditional on the likelihood of machine failure as shown in Figure 4.1. The probabilities of a major breakdown elicited from the implement dealers ranged from 5 to 50 percent over the machine's expected life. These probabilities constitute a base set and can be parameterized to gauge the sensitivity of the optimal policy to changing expectations of machine failure. Each probability in the set is multiplied by a weighting factor whose value ranges from 0 to 200 percent. A factor of 0 percent implies the chances of failure are nonexistent, whereas a factor of 200 percent depicts a failure rate that is twice as high as the base scenario. As the graph illustrates, present costs of an optimal policy are directly related to expected failure rates.

(1,000 dollars)

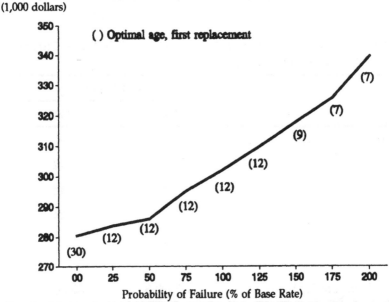

Figure 4.1. Impact of Machine Failure on Optimal Replacement Age and Cost

To mitigate costs, machines are replaced at shorter intervals since failure rates are an increasing function of age. A similar relationship exists with complete equity financing.

Unlike previous studies, alternative marginal tax rates and depreciation policies had minimal effects on optimal replacement strategies. The direction of change was in line with prior studies. Factors limiting the importance of tax policy in this analysis are the low marginal tax rates, repeal of investment tax credit, longer recovery periods and required depreciation recapture provisions of the Tax Reform Act of 1986. Hence, replacement is based upon the true economic and productive merits of a new machine with minimal distortions caused by tax policy.[4]

Involuntary Replacement

The costs of premature trading discussed above could be incurred if the existing asset is involuntarily replaced as a result of an unexpected failure. Whether a failed combine is repaired and retained or traded for a new machine depends in part on the opportunity cost and amount of downtime associated with the breakdown and the degree to which the price of a machine with a known failure is discounted.

With the trade-in value of a failed machine equal to 95 percent of actual value less costs of repair, the optimal decision in this analysis is generally to repair rather than trade a failed machine. The costs of premature replacement and discount applied to a failed machine exceed the opportunity cost of leasing a temporary combine during the time an existing machine is repaired. Involuntary replacement does occur in 36 percent of all possible situations--those near the period when voluntary trading would naturally occur. The sensitivity of the decision rule to changes in these variables is illustrated in Table 4.2. As the value of a failed machine is more deeply discounted and opportunity costs of repairing it diminish, the likelihood of trading declines.

Table 4.2. Likelihood of Involuntary Replacement

Level of Discount Applied % Failed Machine	Opportunity Cost of Down Time		
	$0	$75	$100
	---------percent---------		
0%	43	69	72
5%	0	58	72
10%	0	51	62

Internal Credit Rationing

A credit constraint, imposed either internally or externally, increases the present cost of an optimal policy *ceteris paribus* because the optimal timing of replacement is delayed.[7]   Under external rationing, the firm is unable to replace a machine when it desires because of insufficient fund availability. Of particular interest in this study are the rewards and management responses that accrue from internal rationing.

Rationing may allow the firm to take advantage of unforeseen investment opportunities. The surface of Figure 4.2 illustrates the necessary reduction in purchase price that would be required before various degrees of constraint are self-imposed. The graph shows rather large degrees of constraint have only a minor impact on the value of an optimal policy. This occurs because costs of suboptimal trading are slight over a broad range of initial values.

As a result, small price discounts induce large changes in replacement patterns. In this analysis, a firm would be willing to forego normal replacement 33 percent of the time if it could realize a 2 percent discount on the purchase price of a new combine.

Internal capital rationing can also reduce the costs and need for external debt financing. Rather than invest in an optimal manner and incur financing charges, a firm may delay replacement until sufficient internal funding is available--at the expense of suboptimal replacement. However, as in the previous example, the firm can tolerate considerable delays in replacement before debt financing becomes cost effective because costs of suboptimal trading are minimal (Figure 4.3).

Conclusions

This study develops a combine replacement strategy for a western Minnesota cash grain farm. The decision is based on a function of observed and unobserved state variables and is the solution to a stochastic dynamic programming model which formalizes the tradeoffs between cost minimization, maintenance of cash balances and limitation of unanticipated downtime. Conditions under which both planned and involuntary replacement occur are described. The rewards and management responses that accrue from internal capital rationing are also discussed.

Results of this study have important implications for aggregate investment behavior studies which base replacement on a proportional decay of a continuous capital stock. This fairly smooth representation is actually the sum of many individual binary decisions. Thus, if aggregate investment behavior is to be adequately understood, evaluated, and forecasted, an understanding of important factors influencing individual firm decisions is required.

Value of Optimal Strategy
(1,000 dollars)

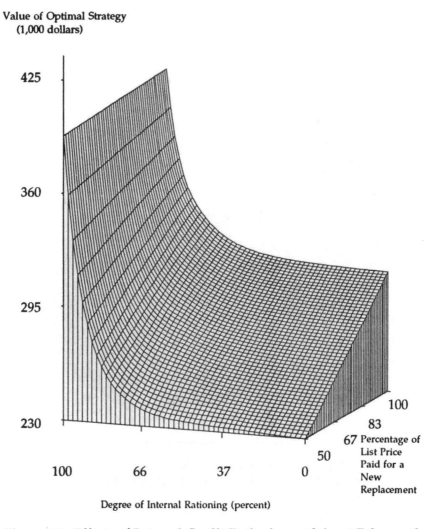

Figure 4.2. Effects of Internal Credit Rationing and Asset Price on the Value of an Optimal Combine Replacement Strategy

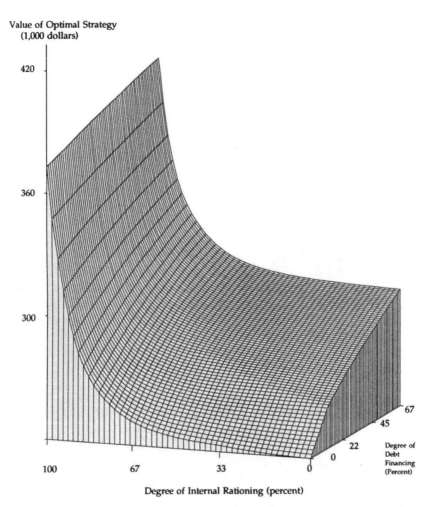

Value of Optimal Strategy
(1,000 dollars)

420

360

300

67

45

22

0

Degree of
Debt
Financing
(Percent)

100   67   33   0

Degree of Internal Rationing (percent)

Figure 4.3. Effects of Internal Credit Rationing and Usage of Debt Capital Value of an Optimal Combine Replacement Strategy

Notes

1. A finite planning horizon explicitly recognizes the life cycle of farms which are predominately organized as sole proprietorships and partnerships.

2. If the replacement asset is financed entirely with equity capital, *DOWN* equals new purchase price and *RBAL* is zero.

3. Even though the expected present cost criterion implies lower total costs with replacement, cashflow required to trade exceeds that of keeping the existing machine because full costs of trading occur in the present stage while benefits accrue in both present and future stages.

4. The need to maintain cash reserves and liquidity maybe internally imposed to meet known transactions demands, done as a precautionary measure to meet unknown adverse needs, or implemented for speculative purposes. In addition, lender and outside equity holders may externally impose capital rationing. A more elaborate cashflow constraint would evaluate the decision with the internal rate of return rather than present cost criterion and specify the distribution of available investment funds, include an additional state variable representing their level at any given stage, permit intertemporal saving, and identify other investment alternatives within the firm. This implies revenues from both the existing and replacement machines are quantifiable.

5. Three dealers represented one combine manufacturer, two represented a second brand while the sixth dealer sold a third brand.

6. This new environment can be challenging as one farmer-member of the panel responded, "Without tax considerations, I no longer know when to replace!"

7. Although a firm's discount rate would be expected to rise under capital rationing, efforts to empirically establish such a linkage proved fruitless.

References

American Society of Agricultural Engineers. *Agricultural Engineers Yearbook.* St. Joseph MO, 1981.

Bates, J.M., A.J. Rayner, and P.R. Custance. "Inflation and Farm Tractor Replacement in the U.S.: A Simulation Model." *American Journal of Agricultural Economics* 61(1979):331-4.

Bradford G. and D. Reid. "Theoretical and Empirical Problems in Modeling Optimal Replacement of Farm Machines." *Southern Journal of Agricultural Economics* 14(1982):109-16.

Burt, O.R. "Dynamic Programming Models With Risk Oriented Criterion Functions." *Risk Analysis For Agricultural Production Firms: Concepts, Informational Requirements and Policy Issues.* Proceedings of Southern Regional Project S-180, Department of Economics and Business, North Carolina State University, Raleigh, July, 1988, p.29-44.

Burt, O.R. "Optimal Replacement Under Risk." *Journal of Farm Economics* 47(1965):324-46.

Chisholm, A. "Effects of Tax Depreciation Policy and Investment Incentives on Optimal Equipment Replacement Decisions." *American Journal of Agricultural Economics* 56(1974):776-83.

Dreyfus, S.E. and A.M. Law. *The Art and Theory of Dynamic Programming*. New York: Academic Press, 1977.

Hardesty, S.D. and H.F. Carmen. *A Case Study of California Farm Machinery: Repair Costs and Downtime*. Giannini Information Series No. 88-2, University of California, Division of Agriculture and Natural Resources, Oakland, May 1988.

Kay, R.D. and E. Rister. "Effects of Tax Depreciation Policy and Investment Incentives on Optimal Equipment Replacement Decisions: Comment." *American Journal of Agricultural Economics* 58(1976):355-8.

Leatham, D.J. and T.G. Baker. "Empirical Estimates of Inflation on Salvage Values, Cost and Optimal Replacement of Tractors and Combines." *North Central Journal of Agricultural Economics* 3(1981):109-17.

Perrin, R.K. "Asset Replacement Principles." *American Journal of Agricultural Economics* 54(1972):60-7.

Reid, D.W. and G.L. Bradford. "On Optimal Replacement of Farm Tractors." *American Journal of Agricultural Economics* 65(1983):326-31.

Spengler, V. *Farm Business Management, 1987 Annual Report*. Northwestern Report No. 33, Thief River Falls Technical Institute, Thief River Falls, MN, April, 1988.

Weersink, A. and S. Stauber. "Optimal Replacement Interval and Depreciation Method for a Grain Combine." *Western Journal of Agricultural Economics* 13(1988):18-28.

5

Optimal Crop Rotations to Control Cephalosporium Stripe in Winter Wheat

Joan Danielson

Cephalosporium stripe is a fungal vascular disease of winter wheat caused by the soil borne pathogen *Cephalosporium gramineum*. Infested winter wheat plants produce fewer and smaller kernels than healthy plants, resulting in significant yield losses. Once an area is infested, control of the disease is important because the fungus survives for long periods of time in winter wheat residues, enabling it to affect subsequent winter wheat crops.

Rotating to non-host spring crops or fallow is the primary method of controlling the disease. This allows time for residue decomposition before winter wheat is seeded again. The number of years of fallow and spring crops between winter wheat crops, together with the Cephalosporium stripe infection level in the last winter wheat crop, determine the level of infection in the current winter wheat crop.

Crop rotation and Cephalosporium stripe infestations are both dynamic processes. As such, current decisions have impacts on future conditions. For example, if the current decision is fallow, the immediate returns will be negative (the variable cost of fallow); but if a crop is planted in the next growing season, yields are generally improved. In addition, fallow adds to the years between winter wheat crops, so future Cephalosporium stripe infection levels will also be affected. The decision to plant a non-host spring crop will usually yield positive immediate returns, but if a crop is planted in the succeeding year, yields will generally be lower than if fallow had been chosen. Spring crops also add a year to the control process so that future winter wheat yields will be affected through future Cephalosporium stripe infection levels. A winter wheat decision has higher expected immediate returns, but has future repercussions on Cephalosporium stripe infestations by returning the system to zero years of control. The system is also put in a continuous-crop versus a fallow-crop situation if the next decision is a spring crop.

An additional complicating factor is the stochastic nature of product prices and Cephalosporium stripe infection levels. At the time of decision making, the magnitude of these variables is not known with certainty. Therefore revenues will not be known with certainty. Accordingly, the problem must be viewed as stochastic, and returns should be viewed in an expected value framework.

In summary, the optimal strategy for control of Cephalosporium stripe infestations in winter wheat involves determination of the sequence of decisions regarding cropping or fallowing the land, such that the expected present value of net returns is maximized. A decision specifies one of the possible alternatives, given expected economic and physical conditions at a particular point in time. A stochastic dynamic problem such as this is efficiently handled by dynamic programming.

The remainder of the paper is organized as follows: the next section develops a general model for Cephalosporium stripe dynamics and winter wheat yield relationships. This is followed by the general decision model and recursive equation. Then the empirical model and results are presented.

The Cephalosporium Stripe Model

Data

In order to formulate a decision model for the control of Cephalo-sporium stripe, relationships concerning infection levels between winter wheat crops, and infection levels and yield must be specified. A study by Mathre et al. (1977) on alternative crop rotations and their effect on Cephalosporium stripe longevity and yield loss at the Moccasin Experiment Station in Judith Basin, Montana provided some of the necessary data. The reduction in winter wheat yields due to infected plants was estimated with this data. The differentials in yields for winter wheat produced on stubble and fallow were estimated from unpublished data for Judith Basin County from the Montana Crop Reporting Service.

The data used to develop and estimate the relationship between initial infection levels and infection levels in subsequent winter wheat crops was output generated from a simulation model developed by plant pathologists at Montana State University (CEPHLOSS).

Cephalosporium Stripe Dynamics

The DP model must consider the effect of the number of years of control on future infection levels. As shown in equation (1), the level of infection is assumed to depend on the level of infection in the last winter wheat crop, the number of years of control, and a random error component.

(1) $I_{(t+1)+C} = f(I_t, C, \varepsilon_{(t+1)+C})$ $C = 1,2,3,...N$

where
I_t = the percent of infected plants in a winter wheat field at time t;
C = the number of years of control (years of fallow and/or nonhost spring crops between winter wheat crops) ;
$I_{(t+1)+C}$ = the percent of infected plants in the same field the next time winter wheat is planted; and
$\varepsilon_{(t+1)+C}$ = the random error term.

The CEPHLOSS output indicates that infection does not begin to decrease until there has been three years of control. After three years of control a type of discontinuous geometric decline begins as years of control increase.[1] A single equation that exhibits a continuous geometric decline in infection with increasing years of control is assumed to be a reasonable approximation of this relationship.

Based on these considerations, the following equation was selected for the Cephalosporium stripe model:

(2) $I_{(t+1)+C} = \exp(\alpha + \beta C + \varepsilon_{(t+1)+C}) * I_t$

Yield-Infection Relationship

Winter wheat yields are assumed to depend on the previous land use (fallow or spring crop), and the level of Cephalosporium stripe infestation in the current winter wheat crop. Previous land use is included to reflect the higher yields, other things equal, associated with winter wheat crops planted on fallowed land versus crops planted on spring crop stubble. An additional assumption is that a given level of Cephalosporium stripe infestation will result in the same percentage reduction in yield irrespective of the previous land use (spring crop or fallow).

Equations (3) and (4) illustrate these hypothesized relationships.

(3) $YF_t = \gamma(1 - bI_t) + \varepsilon_t$

(4) $YS_t = \gamma'(1 - bI_t) + \eta_t$

where

YF_t = winter wheat yield after fallow in growing season t (bushels/acre);

YS_t = winter wheat yield after a spring crop in growing season t (bushels/acre);

γ = winter wheat yield in the absence of the Cephalosporium stripe fungus, given the previous land use is fallow (bushels/acre);

γ' = winter wheat yield in the absence of the Cephalosporium stripe fungus, given the previous land use is a spring crop (bushels/acre);

I_t = the proportion of Cephalosporium stripe infected plants per acre in growing season t;

b = the percentage rate of yield loss due to higher infection levels;

ε_t = a random error term; and

η_t = a random error term.

The General Decision Model and Recursive Equation

The decision alternatives regarding land use in the DP model are winter wheat, fallow, and barley. Fallow and barley represent the methods of control. Barley was chosen as a nonhost spring crop based on historical cropping patterns in Judith Basin County. A 25-year planning horizon is assumed with a decision required annually at winter wheat seeding time in the fall.

Barley seeding time in the spring is another potential decision point. It is not included in this model. A decision to fallow in the fall negates the consideration of planting barley in the spring.

The state variables in the Cephalosporium stripe problem must describe both the economic and physical conditions that will be encountered in a given stage. Winter wheat price and barley price are obvious choices to describe the economic conditions. In addition, winter wheat yields will be affected by previous land use, years of control, and the level of Cephalosporium stripe infection in the last winter wheat crop, all of which are dependent upon past decisions. Spring crops are not subject to Cephalosporium stripe infestations, so that barley yield will depend on the previous land use only. Since infection levels and product prices cannot be predicted with certainty, they are stochastic state variables in the model. Years of control and previous land use are deterministic.

The objective function of the Cephalosporium stripe DP model maximizes the expected present value of net returns from producing winter wheat and barley, subject to the constraints imposed by the state variables. Equation (5) gives the recursive relationship of the DP model.

$$(5) \quad V_t(L_t,C_t,I_t,PB_t,PW_t) = \underset{k}{\text{Max}} \; q^k(L_t,C_t,I_t,PB_t,PW_t)$$

$$+ \beta E \; V_{t+1}(L_{t+1},C_{t+1},I_{t+1},PB_{t+1},PW_{t+1})$$

where

$V_t(\bullet)$ = the maximum expected present value of net returns from year t to the end of the planning horizon (T=25);

L_t = the previous land use;

C_t = the years of control;

I_t = the infection level in the last winter wheat crop;

PB_t = the price of barley;

PW_t = the winter wheat price;

k = the decision variable;

$q^k(\bullet)$ = the expected immediate returns associated with the kth decision alternative (expected total revenue - variable cost);

β = the discount factor, $1/(1.045)$ and

E = the expectation operator.

It is assumed here that the state variables will not affect the value of the firm's assets at the end of the planning horizon. Thus, if t = 25, $V_{t+1}(\bullet)$ equals zero.

The Empirical Model

The data used in the formulation of the Cephalosporium stripe empirical model are representative of Judith Basin County in Central Montana. Cephalosporium stripe infections vary across fields; thus the analysis takes place on a field-specific basis rather than a whole farm basis.

Decision Alternatives

Some restrictions were placed on the decision variable. Two cropping sequences, fallow-fallow and winter wheat-winter wheat, were excluded *a priori* from the model. The soils at the experimental site on which this study is based, are unusually shallow--about 20 inches of soil above a gravel substrata. Therefore, there is seldom any advantage to a second year of fallow (for soil moisture management) because the soil profile is nearly always at field capacity after one season of summer fallow (Burt and Stauber, 1977). The exclusion of two consecutive years of winter wheat is more difficult to defend. However, the Moccasin study does not include data on this rotation and the Cephalosporium stripe infection data derived earlier are based on a worst case scenario which probably would preclude two years of winter wheat.[2]

A third constraint forces a winter wheat decision after six years of control. It is felt that this constraint would not substantially bias the final results and was necessary to limit the number of states associated with Cephalosporium stripe infection levels and years of control.[3]

Given these constraints, the decision alternatives are summarized in Table 5.1.

Transformation Functions

Previous Land Use. Previous land use is a deterministic state variable, assuming a value of 0, 1 or 2, indicating fallow, winter wheat or barley, respectively. Its transition from stage to stage is dependent upon the decision in the previous stage, as shown in equation (6).

Table 5.1 Decision Alternatives of the Empirical Model, Given Previous Land Use and Years of Control

Previous Land Use	Years of Control	Decision Alternative, k
Fallow	1-5	Winter Wheat (1) Barley (2)
	6	Winter Wheat (1)
Winter Wheat	0	Fallow (0) Barley (2)
Barley	1-5	Fallow (2) Winter Wheat (1) Barley (2)
	6	Winter Wheat

(6) $L_{t+1} = k_t, k_t = 0,1,2$

Years of Control. Years of control is a deterministic state variable that denotes the number of years between winter wheat crops. Barley and fallow add a year to control. Winter wheat returns years of control to zero. The transformation function is specified as follows:

(7) $C_{t+1} = C_t + 1,$ if $k_t = 0, 2$ and $C_t \leq 5$
     $= 0,$ if $k_t = 1$

Infection Level in Last Winter Wheat Crop. In defining the transformation function for the infection level variable, it is important to note that I_t, by itself, does not define the expected level of infection in the current time period. It is the level of infection in the last winter wheat crop, a stochastic variable that changes in value from stage t to stage t+1, only if the current decision is winter wheat. Given a winter wheat decision in stage t, the transformation function in general form is given by equation (2). This equation was converted to natural logs for estimation. Then using the generated observations, application of OLS yielded the following coefficients:

(8) $\ln(I_{t+1}) = 1.3210 - 0.76002 * C_t + \ln(I_t), \quad k_t = 1$

If the decision is not winter wheat, a year is added to control, but the level of infection in the last winter wheat crop does not change and is simply carried along to the next stage. Thus the transformation function, as shown below, is not stochastic when the decision is barley or fallow.

(9) $I_{t+1} = I_t, \quad k_t = 0,2$

Winter Wheat and Barley Prices. The transformation functions for the two price state variables formulate the relationship between price in year t and year t+1. The two equations were estimated using 33 years (1951-1983) of Montana wheat and barley prices expressed in 1984 dollars. Because time series data tend to have positive correlation in the successive error terms, both equations were modeled with autoregressive error structures. A zero-one variable was included in the equations to account for the high prices observed in 1973.

The winter wheat equation is a second order autoregressive model. Prices are expressed in natural logs. Equation (10) gives the estimated coefficients with t-values in parentheses.

$$\ln(PW_{t+1}) = \underset{(17.3540)}{1.5462} + \underset{(3.0287)}{0.2892D_t} + \mu_{t+1}$$

(10)

$$\mu_{t+1} = \underset{(7.2055)}{1.1509\mu_t} - \underset{(-2.8630)}{0.4573\mu_{t-1}} + \varepsilon_{t+1}$$

where

$\ln(PW_{t+1})$ = winter wheat price in year t+1 in natural logs;

D_t = the binary variable; $D_t = \begin{cases} 1, \textit{ for } 1973 \\ 0, \textit{ elsewhere} \end{cases}$; and

μ_{t+1} = the second order autoregressive error.

The adjusted R squared was 0.7075 and the standard error is 0.152 in natural logs.

Adjustment of the equation for the autoregressive error structure and setting $D_t=0$ (for 1984) gives equation (11).

(11) $\ln(PW_{t+1}) = 0.4737 + 1.1509 \ln(PW_t) - 0.4573 \ln(PW_{t-1})$

This equation has two lagged values of the dependent variable, indicating the need for two winter wheat price state variables to describe the decision process from stage to stage t+1. To eliminate the need for the additional state variable, the method of reducing the order of a Markov process described by Taylor and Burt (1984) was applied to equation (11). While some information is lost and the conditional variance of PW_{t+1} given PW_t is increased by the modification, these drawbacks are felt to be preferable to increasing the dimension of the problem. The reduction of the lag gives equation (12),

(12) $\ln(PW_{t+1}) = 0.3251 + 0.7897 \ln(PW_t)$

with variance = 0.0292 and standard deviation = 0.171. Note that this standard deviation is not all that much larger than that for equation (10), i.e., 0.152.

The barley price equation is a first order nonstochastic difference equation with a first order autoregressive error structure. Prices are expressed in natural logs. Equation (13) gives the estimation results:

(13) $\ln(PB_{t+1}) = \underset{(2.0146)}{0.20057} + \underset{(4.7735)}{0.56492D_t} + \underset{(7.8858)}{0.77578} E(\ln(PB_t)) + \mu^b_{t+1}$

$$\mu^b_{t+1} = \underset{(2.3904)}{0.38924\mu_t} + \varepsilon_{t+1}$$

where

$\ln(PB_{t+1})$ = barley price in year t+1 in natural logs;

D_t = the binary variable; $D_t = \begin{cases} 1, \text{ for 1973} \\ 0, \text{ elsewhere} \end{cases}$;

$E(\ln(PB_t))$ = the nonstochastic difference equation term; and

μ^b_{t+1} = the first order autoregressive error.

The adjusted R squared is 0.6721 and the standard error in natural logs is 0.125. The model in equation (13) is a discrete intervention type with the once and for all shock of 1973 being dissipated according to geometric decay over the period after 1973, i.e., a geometric distributed lag effect on the 1973 dummy variable.

Equation (13) is adjusted for the nonstochastic difference equation term and autoregressive error term along with setting $D_t=0$ for 1984 resulting in equation (14).

(14) $\ln(PB_{t+1}) = 0.54634 + 0.38924 \ln(PB_t)$

Transition Probabilities

The stochastic state variables, Cephalosporium stripe infection level, winter wheat price, and barley price, are continuous random variables for which probability density functions must be calculated. To facilitate computations, these continuous random variables are made discrete by specifying an arbitrary number of discrete intervals, with midpoint values assigned to represent the associated state intervals. The transition probabilities are calculated by finding the probability of being within the upper and lower bound of an interval (state) identified by its associated midpoint.

The boundaries of the intervals for infection levels (percentages) are 0, 5, 15, 25,....75, 85, 90 with midpoints calculated as the simple average of the upper and lower boundaries of a given interval. Winter wheat and barley price ranges were centered around the long run equilibrium of equations (10) and (13) respectively, with endpoints approximately three standard deviations from the long run mean. Winter wheat prices range from $2.77 through $7.94, and barley prices from $1.69 to $3.54 (midpoint values).

Computations for the transition probabilities were based on equations (8), (12), and (14). Tables 5.2 and 5.3 give the price transition probabilities. Table 5.4 illustrates the transition probabilities for infection given three years of control.[4]

To implement the expectation operator in the recursive equation, a joint transition probability function is needed. Calculation of the joint transition probability function is determined by the relationship between the stochastic state variables. If these variables are mutually independent, the joint transition probability function can be derived by multiplication of the three individual transition probability functions.

Table 5.2 Transition Probabilities for the Winter Wheat Price State Variable

PW_t	PW_{t+1}						
	2.77	3.36	3.97	4.69	5.55	6.56	7.94
2.77	.4989	.3199	.1482	.0303	.0027	.0001	.0000
3.36	.1775	.3353	.3261	.1355	.0238	.0018	.0000
3.97	.0486	.1929	.6591	.2867	.0977	.0141	.0009
4.69	.0084	.0672	.2407	.3675	.2407	.0672	.0084
5.55	.0009	.0141	.0977	.2867	.3591	.1929	.0486
6.56	.0000	.0018	.0238	.1355	.3261	.3353	.1775
7.94	.0000	.0001	.0027	.0303	.1482	.3199	.4989

Table 5.3 Transition Probabilities for the Barley Price State Variable

PB_t	PB_{t+1}						
	1.69	1.87	2.14	2.45	2.80	3.19	3.54
1.69	.0680	.2659	.4023	.2187	.0422	.0029	.0001
1.87	.0296	.1749	.3884	.3097	.0882	.0089	.0003
2.14	.0116	.1017	.3275	.3780	.1568	.0231	.0012
2.45	.0040	.0519	.2423	.4036	.2423	.0519	.0040
2.80	.0012	.0231	.1568	.3780	.3276	.1017	.0116
3.19	.0003	.0089	.0886	.3097	.3884	.1749	.0296
3.54	.0001	.0029	.0422	.2186	.4023	.2660	.0680

Table 5.4 Transition Probabilities for Cephalosporium Stripe Infection Level, Given Three Years of Control

I_t	I_{t+1}									
	2.5	10	20	30	40	50	60	70	80	87.5
2.5	.9871	.0129	.0000	.0000	.0000	.0000	.0000	.0000	.0000	.0000
10	.5892	.3992	.0115	.0001	.0000	.0000	.0000	.0000	.0000	.0000
20	.1635	.6439	.1754	.0164	.0008	.0000	.0000	.0000	.0000	.0000
30	.0212	.5421	.3023	.1119	.0199	.0024	.0002	.0000	.0000	.0000
40	.0015	.3940	.3001	.1954	.0822	.0216	.0043	.0008	.0001	.0000
50	.0001	.2581	.3026	.2048	.1391	.0650	.0221	.0062	.0015	.0005
60	.0000	.1484	.3080	.1953	.1560	.1055	.0538	.0219	.0077	.0035
70	.0000	.0747	.2912	.1942	.1516	.1235	.0837	.0459	.0213	.0139
80	.0000	.0336	.2497	.2007	.1452	.1246	.1004	.0688	.0401	.0370
87.5	.0000	.0170	.2095	.2060	.1433	.1212	.1048	.0810	.0541	.0631

It is easy to argue that the transition probability function for past Cephalosporium stripe infection levels is independent of the price variables. But the same cannot be said for the independence of winter wheat and barley prices. To test for independence of these variables, the winter wheat residuals from equation (10) were regressed on the barley residuals from equation (13). The estimated coefficient on the independent variable (barley residuals) was 0.32320, with a t-value of 1.4554, indicating the price series were, at most, weakly correlated.

Noting that the past infection level variable is only stochastic when the decision is winter wheat, and given the mutual independence of the stochastic state variables, the joint transition probabilities[5] are calculated in the following manner:

$$
(15) \qquad P_{ij} = \begin{cases} \dot{p} \cdot \hat{p} \cdot \bar{p} & , k_t = 1 \\ \hat{p} \cdot \bar{p} & , k_t = 0,2 \end{cases}
$$

where

P_{ij} = the probability of moving from the ith vector of state variables in stage t to the jth vector of state variables in stage t+1;

$\dot{p}$ = the probability of moving from i^{th} infection state to the jth infection state;

$\hat{p}$ = the probability of moving from the ith winter wheat price state to the jth winter wheat price state; and

$\bar{p}$ = the probability of moving from the ith barley price state to the jth barley price state.

Expected Immediate Returns

The expected immediate returns are defined in general as expected total revenue minus variable cost on a per-acre basis. Variable costs of crop production and fallow for a representative Judith Basin dryland grain farm were obtained from Montana State University Cooperative Extension Service costs and returns publications. Since prices and infection levels are statistically independent, their expectations can be calculated and used to derive the expected gross revenue.

A conditional expectation of product price is used in determining the expected immediate returns because there is approximately a one year lag between the decision point (winter wheat seeding time) and the receipt of any revenue from the crop. The state variables, PW_t and PB_t, give the current winter wheat and barley price, respectively. With this information, expectations of winter wheat and barley prices at marketing are formed. These expectations, though formed and implemented in stage t, are actually the conditional expectations of PW_{t+1} and PB_{t+1}.

If the decision in stage t is winter wheat, an additional source of uncertainty, expected infection level, enters the expected immediate returns function. This expectation is conditional on the years of control and the level of Cephalosporium stripe in the last winter wheat crop. The expectation of the infection level in a winter wheat crop in stage t is the conditional expectation of I_{t+1}. That is, in stage t+1, the level of infection in the *last* winter wheat crop (I_{t+1}) would be equal to the actual level of infection in the winter wheat crop experienced in stage t.

Equations (16) and (17) give the parameters for the winter wheat yield-infection relationship discussed earlier (equations (3) and (4)).

(16) $YF_t = 35.3 (1 - .605 (I_t))$

(17) $YC_t = 26.2 (1 - .605 (I_t))$

These equations indicate that if infection is 100% ($I_t = 1$), there will be approximately a 60% reduction in winter wheat yield.

T-ratios and R-squares are not given because the coefficients in these equations are the result of estimation from two different data sets. The parameter b, (.605), was estimated with 18 observations from the Moccasin experiment. But the previous land use for all observations in that data was fallow, so γ' (26.2) could not be determined. As a result, trend lines for winter wheat yields after fallow and continuously cropped were estimated to give the parameters γ (35.3) and γ' from equations (3) and (4).[6]

Trend lines were also estimated to obtain yield figures for barley on fallow (40.6 bu/A) and continuously cropped (36.9 bu/A). All trend lines were estimated with data pertaining to Judith Basin County from the Montana Crop Reporting Service.

Integrating the information on variable cost, expected prices and yields, equation (18) gives the expected immediate returns function associated with each of the decision alternatives and previous land uses. The -\$17.83 for a fallow decision is the negative of the variable cost of fallow, and the bar over the price and infection variables indicates conditional expectations.

$$
(18) \quad q^k =
\begin{cases}
-17.83, & k = 0 \\
\bar{P}^W_{t+1} \cdot (35.3\,(1-.605\cdot \bar{I}_{t+1})) - 59.39, & k=1,\ L_t=0 \\
\bar{P}^W_{t+1} \cdot (26.2\,(1-1.605\cdot \bar{I}_{t+1})) - 72.17, & k=1,\ L_t=2 \\
\bar{P}^B_{t+1} \cdot (40.6) - 53.90, & k=2,\ L_t=0 \\
\bar{P}^B_{t+1} \cdot (36.9) - 78.64, & k=2,\ L_t=1,2
\end{cases}
$$

Results

Solution of the recursive equation yields the optimal policy and expected present value of net returns for all combinations of states and stages. There are 6,370 states in each stage of the Cephalosporium stripe decision model. Sections of the optimal policy which illustrate the trade-offs involved in the decision making process are highlighted in this section.

In Tables 5.5, 5.6, and 5.7 the previous land use is fallow. Thus there are only two decision alternatives: winter wheat and barley. Examined separately, each table illustrates a control/no control frontier. For given winter wheat and barley prices, this frontier depends on the years of control and the past stripe infection level. These factors determine the expected level of infection in the current crop, thus affecting winter wheat yields. As expected, the concentration of winter wheat decisions lies in the lower left corner of each table where years of control are highest and past infection levels are lowest.

Comparison of rows in Tables 5.5 and 5.6 demonstrates the effect of a higher winter wheat price, other things equal. A higher winter wheat price makes winter wheat the more profitable decision at higher levels of past infection. For example, in row three of Table 5.5 (three years of control), there are winter wheat decisions at the two lowest infection levels, while in the same row of Table 5.6 there are four winter wheat decisions. This relationship holds, in general, throughout the optimal policy.

Table 5.5 Optimal Policy Given a Previous Land Use of Fallow, a Barley Price of $1.87, and a Winter Wheat Price of $3.36

Years of Control	Infection in Last Winter Wheat Crop (%)									
	2.5	10	20	30	40	50	60	70	80	87.5
1	W	B	B	B	B	B	B	B	B	B
2	W	B	B	B	B	B	B	B	B	B
3	W	W	B	B	B	B	B	B	B	B
4	W	W	W	B	B	B	B	B	B	B
5	W	W	W	W	W	W	W	W	W	W

Table 5.6 Optimal Policy Given a Previous Land Use of Fallow, a Barley Price of $1.87, and a Winter Wheat Price of $4.69

Years of Control	Infection in Last Winter Wheat Crop (%)									
	2.5	10	20	30	40	50	60	70	80	87.5
1	W	B	B	B	B	B	B	B	B	B
2	W	W	B	B	B	B	B	B	B	B
3	W	W	W	W	B	B	B	B	B	B
4	W	W	W	W	W	W	W	W	B	B
5	W	W	W	W	W	W	W	W	W	W

Table 5.7 Optimal Policy Given a Previous Land Use of Fallow, a Barley Price of $2.45, and a Winter Wheat Price of $3.36

Years of Control	Infection in Last Winter Wheat Crop (%)									
	2.5	10	20	30	40	50	60	70	80	87.5
1	B	B	B	B	B	B	B	B	B	B
2	W	B	B	B	B	B	B	B	B	B
3	W	B	B	B	B	B	B	B	B	B
4	W	W	B	B	B	B	B	B	B	B
5	W	W	W	W	B	B	B	B	B	B

Table 5.8 Optimal Policy Given a Previous Land Use of Winter Wheat, Zero Years of Control, and a Winter Wheat Price of $3.36

Barley Price	Infection in Last Winter Wheat Crop (%)									
	2.5	10	20	30	40	50	60	70	80	87.5
$1.69	F	F	F	F	F	F	F	F	F	F
$1.87	F	F	F	F	F	F	F	F	F	F
$2.14	F	F	F	F	F	F	F	F	F	F
$2.45	F	F	F	F	F	F	F	F	F	F
$2.80	B	B	B	B	B	B	B	B	B	B
$3.19	B	B	B	B	B	B	B	B	B	B
$3.54	B	B	B	B	B	B	B	B	B	B

Tables 5.5 and 5.7 demonstrate the trade-offs that occur when barley price increases, *ceteris paribus*. Again, it is not surprising to find more barley decisions at higher barley prices when the other variables are held constant.

Table 5.8 illustrates portions of the optimal policy for a previous land use of winter wheat and, accordingly, zero years of control. The decision alternatives are limited to fallow and barley under these circumstances. Low barley prices coupled with a previous land use of winter wheat partially explain the fallow-barley frontier. Barley does not become optimal until its price reaches $2.80. But since the expected immediate returns of barley are positive at even the lowest barley price in the model and the immediate returns to fallow are negative, the role of the other variables, in a longer run framework, must be evaluated.

In the optimal policy, most of the winter wheat decisions occur at the two lowest infection levels when the years of control are one, two, or three, unless the winter wheat price is in the upper range. Thus, many of the fallow decisions where the given winter wheat price is low to medium, are probably the first of a three-year control sequence (for example, F-B-F). This sequence allows both a barley and a wheat crop to be planted on fallow. Once the barley price becomes high enough, a different type of control sequence (such as B-B-F) may be optimal for a given winter wheat price.

Not all of the fallow decisions in these tables should be interpreted as the beginning of a longer sequence of control. Fallow associated with the two lowest infection levels is most likely indicative of the end of a control sequence. In the optimal policy, all of the barley decisions at the lowest past infection level are eventually replaced by fallow decisions, as the winter wheat price rises. Some of the barley decisions in the 5 to 10 percent past infection level range also change to fallow under these circumstances. The relatively high winter wheat price, together with the low past infection level, prompts a change in order to exploit the increase in yield associated with a fallow-wheat rather than barley-wheat sequence.

The importance of previous land use is shown in Table 5.9. There are more winter wheat decisions, other things equal, when the previous land use is fallow. If the previous land use is barley and there is a high probability of winter wheat in the next period, the current decision is fallow.

It should be noted that in row 1 of Table 5.9, fallow is not one of the decision alternatives. However, assuming the returns (further yield increases) to a second year of fallow are in fact marginal, the decisions should not be biased.

A broad summarization of the optimal decision rule would indicate that winter wheat decisions occurred most often with one, or a combination, of the following circumstances: high winter wheat prices, years of

Table 5.9 Optimal Policy Given Four Years of Control, Winter Wheat Price of $4.69, and Barley Price of $2.45

Previous Land Use	Infection in Last Winter Wheat Crop (%)									
	2.5	10	20	30	40	50	60	70	80	87.5
Fallow	W	W	W	W	W	B	B	B	B	B
Barley	F	F	F	F	F	F	F	F	B	B

control greater than or equal to three, and lower past infection levels. At the beginning of a control sequence defined by low barley prices, and at the end of a control sequence when conditions were favorable for winter wheat in the next decision period, the predominant decision was fallow. Barley was the optimal method of control when there was a good possibility that one or more additional years of control would be required after the current decision. Given these conditions, barley decisions increase as barley prices increase.

The results presented so far were derived with the condition that the decision maker would consider a maximum of six years of control. To compare the effect of maximum years of control on monetary returns in a given situation, additional results were obtained in which the maximum years of control were reduced to one, two, three, and four years, respectively. Each of the different maximum control specifications yields a set of net present values for each state at each stage of the 25-year planning horizon. Given a vector of the state variables, the discounted expected returns associated with each of the specified values for maximum years of control can be compared. The results are an indicator of the economic value of considering the specified values for maximum years of control.

The significance of the difference in monetary returns due to the length of the years of control maximums are more meaningful when presented on an annual basis rather than a discounted net present value for the planning horizon. Therefore, the net present values were amortized to derive the comparable annual return figures in dollars per acre. The annual return information, given a vector of state variable values under a 25-year planning horizon, is presented in Table 5.10.

Looking down each of the columns of Table 5.10, the cost of limiting the number of years between winter wheat crops to one, two, or three throughout the planning horizon is clearly illustrated. For example, if the Cephalosporium stripe infection level in the last winter wheat crop was in the zero to five percent interval (column one), the amortized returns for a maximum one-, two-, and three-year control sequence are $28.47, $40.17, and $42.30 per acre, respectively.

Looking at the same column but comparing the amortized returns to maximum three, four, and six years of control shows gains associated with allowing six-year control sequences. The gains are small when past infection levels are in the lower range, but in the upper range of past infection levels the returns to a maximum six years of control, rather than three or four, are more significant.

Table 5.10   Amortized Returns from Optimal Crop Sequences in
Relation to Maximum Years of Control and Past
Cephalosporium Stripe Infection Levels Under a 25-Year
Planning Horizon, Given a Previous Land Use of Fallow,
One Year of Control, a Barley Price of $2.45, and a Winter
Wheat Price of $4.69

Maximum Years of Control	Infection in Last Winter Wheat Crop (%)									
	2.5	10	20	30	40	50	60	70	80	87.5
1	28.47	20.32	15.96	13.91	12.85	12.33	12.29	12.29	12.29	12.29
2	40.17	35.53	32.19	29.40	27.10	25.33	23.98	22.96	22.16	21.68
3	42.30	39.56	37.92	36.76	35.91	35.08	34.27	33.50	32.80	32.32
4	42.48	39.72	38.30	37.44	36.72	36.12	35.67	35.50	34.95	34.68
5	42.59	39.83	38.53	37.76	37.22	36.80	36.44	36.12	35.84	35.66

A general recommendation from these results would be to follow a
three-year control sequence when past Cephalosporium stripe infections
are low to medium, but to consider longer control sequences when past
infection levels are more severe. These findings are consistent with the
recommendations of the *Montana Small Grain Guide* (1985) which suggests
a minimum of three years without winter wheat when Cephalosporium
stripe has been observed.

Conclusions

The results of this study indicate that the Cephalosporium stripe
problem is not trivial in terms of returns to the winter wheat producer.
However, the quality of the results are data dependent. The most
obvious inadequacy in the model is the lack of experimental data to
statistically estimate the infection relationship. Experimentation designed
to provide such data would improve the reliability of the results.

The results are also region specific, especially with regard to the
method of control. The choice of the spring crop is separable from the
control/no control decision. The effect of fallow on crop yields is largely
determined by soil composition and depth. Two years of fallow may be
optimal in some areas. In others, one year may not have the impact on
yield shown in this study, thus reducing the number of fallow decisions.

The model could be extended to allow for decisions regarding land use
to be made in the spring as well as the fall. In a semi-annual model, the

inclusion of a state variable measuring soil moisture would provide valuable information. It is clear that there is a threshold value for soil moisture for profitable crop production. That is, even if all other conditions indicated a crop should be planted, the level of soil moisture could indicate that the outcome would be uneconomical.

The present annual model would be improved by fall soil moisture measurements when winter wheat was being considered following a summer in barley. However, given the unusually shallow soils at the experimental site, soil moisture in the fall is not informative when the land was in summer fallow. Thus the present model is essentially complete.

Notes

1.   The equations from CEPHLOSS are:

$$
I_{(t+1)+c} = \begin{cases}
2I_t, & C = 1 \\
I_t, & C = 2 \\
(1/3)^{(1/2)(C-1)}I_t, & C = 3,5,7,... \\
(1/3)^{(1/2)(C-2)}(7/9)I_t, & C = 4,6,8,...
\end{cases}
$$

Note the equations do not consider the possibility of continuous winter wheat (C = 0). Infection levels are also restricted to be $\leq 90\%$ in the simulation model.

2.   Donald E. Mathre, personal communication.

3.   The infection transition probabilities, given six years of control and any initial infection level, indicate that there is a very low probability of getting infection levels above 15%.

4.   There are six infection transition probability matrices, one for each year of control considered. A copy of the others is available from the author upon request.

5.   For greater computational detail on either the individual or joint transition probabilities, see Danielson (1987).

6.   For details see Danielson (1987).

References

Bellman, R. *Adaptive Control Processes*. Princeton, NJ: Princeton University Press, 1961.

Bellman, R. *Dynamic Programming*. Princeton, NJ: Princeton University Press, 1957.

Brockus, W.N., J.P. O'Connor, and P.J. Raymond. "Effect of Residue Management Method on Incidence of Cephalosporium Stripe Under Continuous Winter Wheat Production." *Plant Disease* 47,12(1983):1323-24.

Bruehl, G.W. "Cephalosporium Stripe Disease of Wheat." *Phytopathology* 47(1957):641-9.

Burt, O.R. and J.R. Allison. "Farm Management Decisions with Dynamic Programming." *Journal of Farm Economics* 45(1963):121-36.

Burt, O.R. and M.S. Stauber. "Economic Analysis of Cropping Systems in Dryland Farming." Final Report, Old West Commission Project No. 10470025. Agricultural Economics and Economics Department and Montana Agricultural Experiment Station, Montana State University, Bozeman, Montana, April 1977.

Danielson, Joan G. "Optimal Crop Sequences to Control Cephalosporium Stripe in Winter Wheat." Unpublished Master's Thesis, Montana State University, Bozeman, Montana, 1987.

Dreyfus, S.E. and A.M. Law. *The Art and Theory of Dynamic Programming.* New York: Academic Press, 1977.

Fogle, V. "1982 Update: Enterprise Costs for Fallow, Winter Wheat, Barley after Fallow, and Recrop Barley in Judith Basin County." Bulletin 1210 (revised). Montana Wheat Research and Marketing Committee and Montana Cooperative Extension Service, Montana State University, Bozeman, Montana, September 1982.

Howard, R.A. *Dynamic Programming and Markov Process.* New York: John Wiley and MIT Press, 1960.

Johnston, B. and G. Hahn. "CEPHLOSS Computer Program." Montana Agricultural Experiment Station, Montana State University, Bozeman, Montana, 1984.

Latin, R.X., R.W. Harder, and M.V. Wiese. "Incidence of Cephalosporium Stripe as Influenced by Winter Wheat Management Practices." *Plant Disease* 66,3(1982):229-30.

Mathre, D.E., A.L. Dubbs, and R.H. Johnston. "Biological Control of Cephalosporium Stripe of Winter Wheat." Capsule Information Series. Montana Agricultural Experiment Station, Montana State University, Bozeman, Montana, December 1977.

The Montana Small Grain Guide. Bulletin 364. Montana Cooperative Extension Service, Agricultural Experiment Station, Montana State University, Bozeman, MT, August 1985.

Morton, J.B., D.E. Mathre, and R.H. Johnston. "Relation Between Foliar Symptoms and Systemic Advance of *Cephalosporium gramineum* During Winter Wheat Development." *Phytopathology* 70(1980):802-7.

Nisikado, Y., H. Matsumoto, and K. Yamauti. "Studies on a New Cephalosporium, Which Causes the Stripe Disease of Wheat." *Ber. Ohara Insts.*, Landwirtsch Forsch., Kurashiki, Japan 6 (1934):275-306.

Pool, R.A.F. and E.L. Sharp. "Some Environmental and Cultural Factors Affecting Cephalosporium Stripe of Winter Wheat." *Plant Disease Reporter* 53,11(November 1969):898-902.

Stauber, M.S., O.R. Burt, and F. Linse. "An Economic Evaluation of Nitrogen Fertilization of Grasses When Carry-over is Significant." *American Journal of Agricultural Economics* 57,3(1975):463-71.

Taylor, C.R. "A Simple Method for Estimating Empirical Probability Density Functions." Staff Paper No. 81-1. Department of Agricultural Economics and Economics, Montana State University, Bozeman, Montana, January 1981.

Taylor, C.R. and O.R. Burt. "Near-Optimal Management Strategies for Controlling Wild Oats in Spring Wheat." *American Journal of Agricultural Economics* 66,1(1984):50-60.

_____. "Reducing the Order of Markov Processes to Reduce the State Variable Dimension of Stochastic Dynamic Optimization Models." Unpublished research paper. Department of Agricultural Economics and Economics, Montana State University, Bozeman, Montana, 1985.

White, D.J. *Dynamic Programming.* San Francisco, CA: Holden-Day, Inc., 1969.

Zacharias, T.P. and A.H. Grube. "Integrated Pest Management Strategies for Approximately Optimal Control of Corn Rootwork and Soybean Cyst Nematode." *American Journal of Agricultural Economics* 68,3(1986):704-15.

6

Optimal Participation in the Conservation Reserve Program

Celia Ahrens Johnson

In December of 1985, the 99th Congress passed and the President signed into law a farm bill to govern federal agricultural and food policy for the next five years. This paper investigates the impact of this legislation on land retirement decisions made by individual producers. Specifically, a stochastic, dynamic program is presented which models a profit maximizing Illinois producer's decision to enroll highly erodible farmland in the Conservation Reserve Program.

The Conservation Reserve Program

The conservation compliance provision of the 1985 Food Security Act requires that all producers cropping highly erodible land begin implementing a conservation plan by January 1, 1990, in order to be eligible for program benefits; implementation of the plan must be completed by January 1, 1995. The conservation reserve provision of the Act establishes a national reserve of highly erodible land, scheduled to encompass 40-45 million acres, which is approximately 10 percent of the nation's cropland, by 1990 (U.S. Congress 1985).

A producer wishing to participate in the Conservation Reserve Program (CRP) must submit a bid for annual government payments based on the costs associated with removing the land from production and establishing a conservation cover. If the producer's bid is accepted, he[1] is paid half the cost of planting perennial grasses or trees plus an annual fee for conservation maintenance during a ten-year contract term. The plan must be approved by the local conservation district, and the producer must agree not to harvest, graze, or otherwise commercially use the vegetative cover. A producer who returns land to production before his contract expires must repay the government, with interest, all of its contributions to date (Glaser 1986).

Of the conservation techniques available to control soil loss, the CRP represents the most extreme measure: removing land from production and planting perennial vegetation to halt erosion. Generally speaking, the more erosive the land and the lower its productivity, the greater the social benefits from withdrawing that land from crop production. To the producer, however, the net benefits of land retirement are seldom high enough to prompt this action unless the government reimburses him for his conservation efforts. Many studies have found private economic incentives for soil conservation to be weak (see, for example, Seitz and Swanson 1980; Burt 1980; Lee et al. 1975; Narayanan et al. 1974; Swanson and MacCallum 1969). The CRP transfers some of the conservation costs from the producer to the current and future citizens who reap many of the benefits of soil conservation.

The intent of the framers of CRP legislation was that producers would bid competitively for the opportunity to enroll land in the CRP. Under a truly competitive bidding system, a producer's bid would reflect the marginal returns associated with farming his highly erodible cropland. During the initial bidding process, however, the government established bid caps for each county. Once the existence of bid caps became known, producers began bidding the maximum rental rates allowable in their counties.

The lack of a true bidding system makes the rates quite generous for relatively unproductive erodible land, but too low to motivate enrollment of land which is both highly productive and highly erodible. A substantial portion of Illinois cropland eligible for the CRP falls in this latter category. This paper investigates the CRP enrollment decision from the perspective of an Illinois producer assumed to be maximizing expected profits. The producer's optimal CRP enrollment decision rule is derived for two representative Illinois farms using the technique of stochastic dynamic programming.

The Producer's CRP Enrollment Decision

Under the 1985 Food Security Act, enrollment in the CRP takes place between 1986-1990 during sign-up periods scheduled at the discretion of the Secretary of Agriculture. At present, there are no provisions for continuing program enrollments beyond the year 1990. When the Act was signed, a producer had five years to decide whether to enter land in the CRP, and once enrolled, his land remained in the program for ten years unless he agreed to pay the penalty for early withdrawal.

A producer maximizing expected profits would make the decision to enroll a field in the CRP if he expected returns from program participation to exceed net returns from farming the field. Since the contract encompasses ten years, the producer would compare the discounted expected net returns from farming the field during that period with the discounted returns he would receive on the contract rental price. The producer would make a decision to enter a field in the CRP when the following inequality held:

$$\sum_{t=1}^{10} E\left[\left(\frac{1}{1+d}\right)^t R_t\right] < \sum_{t=1}^{10}\left[\left(\frac{1}{1+d}\right)^t C\right]$$

where
    t = year;
    E = expectations operator;
    d = producer's rate for discounting;
    R = net returns from farming; and
    C = contract rental rate.

He would not enter his field if the inequality were reversed.

In the above inequality, the rental rate is a constant amount, while net returns to farming are a function of time. This implies that returns to a CRP contract are certain, while returns to farming are stochastic because they depend upon yield and price uncertainties beyond the producer's control.[2] Yield fluctuates with climate and with biological cycles of pests; price fluctuates with market conditions, government policies, and macroeconomic factors.

The inequality just presented is illustrative but oversimplified. A maximizer of expected profits will evaluate after-tax rather than before-tax returns, and in order to properly evaluate after-tax returns, the producer must look at returns for the whole farm. He cannot calculate after-tax returns for a single field because taxes are a function of total income. Therefore, he must compare the whole-farm, after-tax returns to farming with those to CRP participation for all combinations of eligible fields to determine the profit-maximizing combination of alternatives.

A Farm-Level Model of CRP Entry Decisions

The stochastic, multi-year nature of the producer's CRP enrollment decision process makes his problem particularly amenable to solution by dynamic programming (DP), which describes a problem as a system in terms of stage, state, and decision variables. Each stage (or time period) is characterized by several possible states, which are defined by the values of the various state variables that describe the system. A decision is made at each stage which optimizes an objective function and controls the state of the system during the next stage.

The model developed here is not intended to be a complex replication of the producer's decision process. Rather, the intent is to create a relatively simple model that captures the essential aspects of the process yet is still flexible enough to be operational for investigation of different incentives that might be offered for enrolling land in the CRP.

Underlying Assumptions

The producer's objective is to find the combination of farming and CRP enrollment decisions for his eligible fields that maximizes the present value of expected net returns over his entire planning horizon. The 15-year planning horizon for this problem reflects the fact that the length of the enrollment period (5 years) and the length of the contract term (10 years) are fixed by the 1985 Food Security Act. Each passing year of the enrollment period effectively shortens the producer's planning horizon for CRP participation.

The producer is assumed to own a farm composed of a number of different fields, some of which are highly erodible and thus eligible for enrollment in the CRP, and some of which do not meet CRP entry requirements and thus will always be farmed. For the purposes of this analysis, two representative farms were defined, based on the dissimilar soil composition of two Illinois counties. Farm size of 700 acres remains fixed throughout the planning horizon. Once entered in the CRP, a field remains in for the duration of the contract.[3]

The producer maximizes whole-farm, after-tax returns when making his land-use decisions. His returns include not only current profits, but discounted present value of future returns. His entire income is derived from his farm, either through the sale of crops or from rental payments for acreage enrolled in the CRP. In this analysis, the producer does not participate in federal commodity programs.[4] He does farm his land in compliance with conservation standards established for the State of Illinois' T-by-2000 program, however (Illinois Department of Agriculture 1985a and 1985b).

Basic Variables

Stage Variable. The stage for this model is defined to be a year, since planting decisions must be made annually. The planning horizon thus consists of fifteen stages.

State Variables. Each field of highly erodible land is characterized by two state variables, one stochastic and the other deterministic. The deterministic state variable indicates the number of years that a field has been in the CRP. Values of this variable range from 0 to 10, where 0 denotes farming and 1-10 encompass the possible contract years. A deterministic state transition equation describes the movement between states from one stage to the next.

The stochastic state variable indicates the net returns to farming a field over ten possible returns states. Returns must be treated as stochastic because the requirements for certainty equivalence are not met in this problem.[5] The state transition information for the stochastic returns variable is provided by a farm-specific probability transition matrix rather than by a deterministic transition equation. Since there are ten possible returns states, the 10x10 matrix for each farm consists of 100 probabilities, each indicating the probability of making a transition between a returns state at time t and a returns state at time t+1.

Decision Variables. For each highly erodible field, the producer must make a decision to farm it or to enroll it in the CRP. Thus, there must be a decision variable associated with each field. The field decisions are interdependent, however, because the producer's final set of land-use decisions is based on whole-farm, after-tax returns rather than before-tax returns that can be calculated for individual fields.

Once the decisions have been made, each field's land-use status becomes the starting point for next year's land use decisions. When a field is enrolled in the CRP, the land-use decision is determined for ten years. After the contract expires, the field is farmed once again. Through this annual decision process, the state of the entire system is revealed by the values of the state variables in any given stage. It is not necessary to know decisions from previous stages because those decisions are captured by the values of the state variables in the current stage. This attribute of the decision process satisfies the Markovian requirement of dynamic programming (Kennedy 1981).

Formal Specification

Given the assumptions and definitions discussed above, a complete statement of the producer's objective is captured by the following recursive equation:

$$V_t(R_t, Y_t) = \max_{x_t^i} [T(\sum \pi_t^i) + \beta EV_{t+1}(R_{t+1}, Y_{t+1})]$$

where

    i = field index;

    t = year of planning horizon;

    β = discount factor   ($1 \div 1+d$, where d = discount rate);

V = discounted present value of net returns;

E = expectations operator;

T = tax function for combined federal and state tax rates;[6]

$$\pi = \quad X = \begin{cases} \text{price} \times \text{yield} - \text{costs} \times \text{inputs} & : \text{if decision is to farm field} \\ \text{contract rental price} & : \text{if decision is to enroll field} \end{cases};$$

decision variable:

$$X = \begin{cases} 0: & \text{if decision is to farm field} \\ 1: & \text{if decision is to enroll field} \end{cases};$$

state variables:

R = whole-farm, after tax net returns;

Y = vector of years in CRP (0-10);

state transition equations:

$R_{t+1} = f(R_t, \varepsilon_t)$ next year's whole-farm, after-tax net returns—a stochastic function of this year's returns; and number of years field

$$Y_{t+1} = \begin{pmatrix} 0 : & \text{if decision is to farm field} \\ Y_t + 1 : & \text{if decision is to enroll field} \end{pmatrix} \text{ has been in the CRP.}$$

The objective function states that the producer's decisions maximize the sum of current after-tax returns over all farm fields plus the discounted present value of expected future returns. State transition equations describe the movement of the system from one state to the next. The returns function (R_{t+1}) is stochastic, relying on probabilities to describe the state transition, and the years function (Y_{t+1}) is deterministic. The two state transition equations relate the total decision process at each stage to its adjacent stages. Note that π indicates returns per field, and R indicates whole-farm returns.

Limitations on the Model

The producer's decision consists of selecting one of a number of alternative states, each of which involves a different land-use combination of fields. The number of possible alternatives increases dramatically with the number of eligible fields because the producer must consider all possible combinations of years in the CRP for all eligible fields. Such an evaluation is necessary because with multiple eligible fields there may be an optimal order of entry into the CRP. During the five-year enrollment period, the producer need not make all his CRP contracts in a single year.

He may choose to enroll several fields over the course of different enrollment periods.

The formula for calculating the number of possible alternatives, or states, in a DP problem is a multiplicative function of the number of possibilities associated with stages, state variables, and decisions:

number of stages × number of states × number of decisions

For multiple state variables, the number of states is a multiplicative function of the number of discrete categories associated with each state variable:

variable 1 × variable 2 × variable 3 × variable 4 . . .
categories   categories   categories   categories

The producer's planning horizon comprises 15 stages. There are two state variables, net returns to farming and number of years in the CRP. Farm returns are divided into 10 discrete categories from low to high. The number of years in the CRP ranges from 0 (farming) to 10 (1-10 years of elapsed contract), yielding 11 possible alternatives for the land-use status of a field. The decision alternatives are two: farm the land, or enroll it in the CRP.

Given the assumption that the producer will maximize whole-farm, after-tax returns, the number of CRP-eligible fields becomes the limiting factor in model formulation because of the need to consider all possible combinations of years enrolled for all possible eligible fields. The possible number of states associated with the years variable is thus the number of fields: categories$^{(fields)}$. The more restrictive the model, the more one can constrain the number of possible combinations, and the greater the number of eligible fields that can be included. However, given the exponential relationship between years and fields, the number of eligible fields will necessarily remain fairly small.[7]

The Representative Farms

Soil productivity and erodibility are hypothesized to be influential in CRP enrollment decisions. In an attempt to capture their influences, two 700-acre representative farms were designed, based on Illinois counties dissimilar in both respects. The farms reflect the predominant soil types and overall soil composition of Champaign and Jackson counties.

Champaign County, located in east-central Illinois, is one of the most productive counties in the state. Over 90 percent of its 640,000 acres meets the USDA requirements for prime agricultural land,[8] and erosion is a major problem on only about 4 percent of the county's land (USDA, SCS 1982). Jackson County, comprising 387,200 acres in southwestern Illinois, is much less productive in terms of agricultural output. According to the county soil survey (USDA, SCS 1979), only 68 percent of the total acreage is suitable for growing corn and soybeans, the crops of choice in the area. Soil erosion is widespread, and considered to be a major problem on nearly half of county farmland.

The representative farms are constructed of "fields" based on soils predominant in each county. The farms are not intended to reproduce conditions present on any *particular* Illinois farm; rather, each represents a farm of "typical" county soils. To create the "fields," county soils were divided into two groups, those classified as HEL and those not so classified, and the most prevalent soils in each group were selected to meet the county's percentage of soils in that group. For convenience, all field size calculations were rounded to the nearest ten acres.

Problems associated with averaging different soil characteristics were avoided by having each field consist of a single soil type. Although the representative farms contain only a small number of the total soils in each county, the percentages of highly erodible land are close to the actual percentages in the respective counties: 8 percent of the Champaign farm, and 46 percent of the Jackson farm.

Composition of the two representative farms by soil type and by acreage is detailed in Table 6.1. The Champaign farm consists of six soil types; three fields are eligible[9] for the CRP, and three are not. The Jackson farm consists of eleven soil types--three eligible fields, and eight which are ineligible.[10] Table 6.1 also shows the crop rotations, tillage systems, and conservation practices selected for each field to comply with the 1985 Food Security Act and the State of Illinois' T-by-2000 program. These conservation plans reflect practices approved by the Soil Conservation Service (SCS) and the local Soil and Water Conservation District, as detailed in the Illinois SCS series of Cropland Resource Management Systems (USDA, SCS 1987).

Returns to Farming and to CRP Participation

Net profits on the representative farms are a function of three components: prevailing prices received for crops planted, the yields of those crops, and the costs incurred in their production. Calculating net returns to farming for the representative farms involved several steps: computing the average expected returns per acre for each field using current data, estimating whole-farm returns functions with time series data, and adapting the field-specific returns to the stochastic framework provided by the whole-farm returns functions. The process of calculating returns is described in detail below.

Estimating Field-Specific Returns

For a producer not participating in commodity programs, per acre returns are a straightforward summation of market price times yield less costs over the acres planted to each crop, divided by the total acres planted:

$$PAR = \left[\sum_c A_c \{ MP_c \cdot Y_c - (VC_c + CC) - (FC_c + TC) \} \right] \div \sum_c A_c$$

Table 6.1 Acreage Composition of the Representative Farms

Field	CRP Status	Soil	County Acres	Farm Acres	EI[a]	PI[b]	Rotation[c]	Tillage[d]	Practice[e]
Champaign Farm									
1	Ineligible	Drummer	248094	420	2	150	CS	Conv	None
2	Ineligible	Flanagan	99607	170	2	160	CS	Conv	None
3	Ineligible	Elliott	31039	50	4	130	CS	Conv	Contour
4	Eligible	Catlin	16069	30	7	145	CS	Mixed	Contour
5	Eligible	Varna	11142	20	8	125	CS	Mixed	Contour
6	Eligible	Parr	5821	10	13	105	CS	No-till	
Jackson Farm									
1	Ineligible	Belknap	11565	60	3	115	CS	Conv	None
2	Ineligible	Darwin	10004	50	2	95	CS	Conv	None
3	Ineligible	Wakeland	9388	50	3	142	CS	Conv	None
4	Ineligible	Bonnie	8374	40	2	100	CS	Conv	None
5	Ineligible	Piopolis	5635	30	3	75	CS	Conv	None
6	Ineligible	Cairo	5328	30	1	105	CS	Conv	None
7	Ineligible	Okaw	12728	60	3	80	CS	Conv	Contour
8	Ineligible	Hurst	11847	60	5	90	CS	Conv	Contour
9	Eligible	Alford	13847	70	9	125	CS	Mixed	Contour
10	Eligible	Hosmer	32080	160	13	115	CS	No-till	Contour
11	Eligible	Hosmer	17310	90	40	80	CSWH	No-till	Constrip

[a]EI: Water erosion index, based upon USLE.
[b]PI: Soil productivity index.
[c]CS: corn-soybean rotation; CSWH: corn-soybean-wheat-hay rotation.
[d]Conv: conventional tillage (chisel,disk; residue<30%); Mixed: no-till (corn), mulch-till (soybeans); No-Till: no-till (corn and soybeans).
[e]Contour: contour plowing; ConStrip: contour stripcropping.

where
PAR = per acre returns;
c   = crop index;
A   = acres;
MP = market price;
Y   = average per acre yield for soil type;
VC   = 1986 variable cost;
CC   = cost of contour plowing or contour stripcropping;
FC   = 1986 fixed cost; and
TC   = additional fixed cost for tillage system.

The commodity prices used to calculate field-specific returns were derived from prices reported for relevant commodities in *Illinois Agricultural Statistics* for the period 1961-1986. Commodity prices were adjusted for technological change by adopting simulated equilibrium prices generated by AGSIM (Taylor 1989), a national agricultural policy simulation model:

Corn $2.71/bu
Soybeans   5.41/bu
Wheat   2.78/bu
Hay   67.40/ton

Much of the variability in returns by field can be traced to yield differentials attributable to soil quality.   According to the *Illinois Agronomy Handbook* (CES 1987-88), conventional and conservation tillage produce comparable yields on most Illinois soils when stands are adequate and pests are controlled--both characteristics of the high level of management assumed in this analysis. Reduced yields may occur with conservation tillage on poorly drained soils, but all of the highly erodible fields on the representative farms which require conservation tillage are well-drained to moderately well-drained (USDA, SCS 1982, 1979). Table 6.2 presents the average yield of the relevant commodities on all fields of the Champaign and Jackson farms.

The data used to calculate costs of production for each field are those reported in the Soil Conservation Service's (1986) *Field Office Technical Guide*.[11]   Variable cost figures listed in the *Field Guide* for corn and soybeans differ only slightly for conventional tillage, mixed tillage, and no-till cultivation. The principal cost differences between these tillage systems are manifest in fixed costs. The *Field Guide* also reports variable cost of $3.00 associated with contour plowing and $2.50 for contour stripcropping.

With a typical corn/soybean rotation, corn and soybeans are planted in alternate years. To simplify calculating average annual expected returns, the assumption was made that all crops are planted on all fields every year. For example, on a 100-acre field, 50 acres would be in corn and 50 in soybeans; the following year, the crops would be reversed. Over a period of several years, averaging returns in this manner should give the same result as alternating crops.[12]

Per acre returns to participation in the CRP are based solely on the rental rate offered by the government. The assumption is made that producers, when writing their contract applications, will submit as their rental rate the bid cap approved by the Secretary of Agriculture for their particular county. Bid caps for the 1986-1988 sign-ups were $90/acre for Champaign County and $60/acre for Jackson County. The Secretary of Agriculture approved a maximum of $115/acre for Champaign County and $70/acre for Jackson County, but those counties were still operating with the original bid caps just prior to the sixth enrollment period. Table 6.2 shows the returns to farming and to CRP participation for each field, as well as the average expected yield for each crop planted.

Estimating Whole-Farm Returns

Once per acre returns were determined for each field, the total alternative returns to each highly erodible field could be computed for both farming and for CRP participation. Unfortunately, the average expected field-specific returns do not incorporate uncertainty or the dynamics of planning over time. The returns are based on a single set of prices, yields, and production costs. Uncertainty implies that there are probabilities associated with different levels of returns. Generating these probabilities requires econometric estimation of a whole-farm returns function for each representative farm using time-series data for price, production, and cost. The resulting returns function can then be used to develop a matrix of probabilities associated with the transition from one returns state to another.

Crop-specific data for yields and prices from 1961 to 1986 were available in *Illinois Agricultural Statistics*. Variable cost data were not available for Illinois, but national figures from 1961-1978 were obtained from Gallagher and Green (1984) and for subsequent years from USDA *Cost of Production* surveys. Many of these cost data were specific to the Great Lakes and Corn Belt states. The Gallagher and Green figures were indexed to make them compatible with USDA data from later years.[13] Fixed cost data, which are not crop specific, were obtained from the same sources.

Table 6.2 Per Acre Returns to Farming and to CRP Participation for the Representative Farms

Field	CRP Status	Acres	Rotation	Corn	Beans	Wheat	Hay	Farm	CRP
				Expected Yield[*]				Per Acre Returns	
Champaign Farm									
1	Ineligible	420	CS	154	51	--	--	188	--
2	Ineligible	170	CS	162	5	--	--	199	--
3	Ineligible	50	CS	127	45	--	--	140	--
4	Eligible	30	CS	149	46	--	--	174	90
5	Eligible	20	CS	122	41	--	--	134	90
6	Eligible	10	CS	105	37	--	--	102	90
Average per acre returns to farming								184	
Jackson Farm									
1	Ineligible	60	CS	105	38	--	--	104	--
2	Ineligible	50	CS	90	32	--	--	73	--
3	Ineligible	50	CS	135	47	--	--	158	--
4	Ineligible	40	CS	90	32	--	--	73	--
5	Ineligible	30	CS	90	32	--	--	73	--
6	Ineligible	30	CS	103	36	--	--	95	--
7	Ineligible	60	CS	73	25	--	--	35	--
8	Ineligible	60	CS	76	29	--	--	48	--
9	Eligible	70	CS	120	42	--	--	133	60
10	Eligible	160	CS	105	37	--	--	103	60
11	Eligible	90	CS	75	26	34	2.5	44	60
Average per acre returns to farming								86	

[*]Yield for corn, beans, and wheat in bushels per acre; yield for hay in tons per acre; dash indicates crop is not grown on the field.

Two time-series returns functions were estimated, one for each representative farm, based on district-level price data, county-level yield data, and regional cost of production data. The data were customized to reflect production on the representative farms by specifying the acreage planted to each of the different crops. Annual returns for each farm were calculated using the following formula:

$$R_t = \sum_c [P_{ct} \cdot Y_{ct} - VC_{ct} - FC_{ct}] \cdot A_c$$

where

t	= year index;
c	= crop index;
R	= returns;
P	= market price per bushel;
Y	= yield per acre;
A	= acres planted;
VC	= variable cost per acre; and
FC	= fixed cost per acre.

The resulting net returns time series for each representative farm was then converted to 1986 dollars using the GNP implicit price deflator. Examination of the real returns data showed unusually high per acre returns for the years 1972 and 1973, years associated with high market prices because of the grain sale to the Soviets. Dummy variables constructed for 1972 and 1973 were statistically significant (see Table 6.3), so those years were dummied out of the returns functions on the assumption that such a felicitous grain sale could not be presumed to occur during the period from 1986 to 2000 over which the analysis takes place.

Table 6.3   Results from Econometric Estimation of Whole-Farm Returns for the Representative Farms

Indicator	Champaign Farm	Jackson Farm
Overall		
Adjusted R^2	.7543	.4251
Standard Error of Estimate	43.80	43.74
Coefficients		
β_0	59.91 (2.3)[a]	43.82 (2.6)[a]
β_1	.6497 (5.9)	.3745 (2.2)
β_2	145.29 (3.2)	112.61 2.5
Residuals		
Goodness of Fit	3.45[b]	9.85[b]
Jarque-Bera LM	.226	.249
Descriptives		
Minimum	69.88	12.48
Maximum	368.61	168.90
Mean (excluding 1972, 1973)	206.64	84.27

[a] t-statistic
[b] X^2 statistic

The returns function resulting from econometric estimation was thus specified as

$$R_t = \beta_0 + \beta_1 R_{t-1} + \beta_2 D_{1972} + \beta_3 D_{1973} + \varepsilon_t$$

where

R_t = returns in current period;
R_{t-1} = returns in previous period;
D_{1972} = dummy variable for 1972;
D_{1973} = dummy variable for 1973; and
ε_t = stochastic error in current period.

Table 6.3 presents the estimation results of each equation as well as some basic descriptive statistics concerning per acre returns for each of the representative farms.

Probability Transition Matrices

The econometrically estimated returns functions provide the basis for calculating transition probabilities between returns states for each farm. To define the returns states, each farm was assigned a feasible range of per acre returns, which was then divided into ten equal increments to represent ten possible returns states. Champaign County's greater soil productivity, and hence higher average yields, resulted in a wider range of returns for the Champaign farm ($70-$370) than for the Jackson farm ($12-$170). Hence, the returns intervals are wider for the former ($40) than for the latter ($20). Using these values, a probability transition matrix was constructed for each farm using Taylor's (1984) hyperbolic tangent approximation to a cumulative normal distribution. The dimension of each matrix is 10×10, with each cell indicating the probability associated with going from one revenue level in year t to another revenue level in year t+1.

The probability transition matrices for each farm are presented in Table 6.4. Probabilities in the transition matrix for the Champaign farm tend to be highest in the vicinity of the principal diagonal, while those for the Jackson farm are more widely distributed. This can be attributed to the fact that much more of the variation in returns was explained for the Champaign farm (adjusted R^2=.7543) than for the Jackson farm (adjusted R^2=.4251). The matrix for the Jackson farm suggests that county producers have greater difficulty in estimating the state of next year's returns than their Champaign county counterparts due to the greater variability in yield associated with farming lower quality soil.

Table 6.4 Probability Transition Matrices for the Representative Farms

Revenue Level, Year t	Revenue Level, Year t+1									

CHAMPAIGN FARM

-------------------------------------Mean Value of Returns----------------------------------

	$70	110	150	190	230	270	310	350	390	430
$70	.301	.351	.252	.083	.012	.001	.000	.000	.000	.000
110	.130	.286	.342	.188	.047	.005	.000	.000	.000	.000
150	.042	.166	.332	.305	.129	.025	.002	.000	.000	.000
190	.010	.068	.229	.351	.248	.080	.019	.001	.000	.000
230	.002	.020	.113	.289	.341	.185	.046	.005	.000	.000
270	.000	.004	.039	.170	.334	.302	.125	.024	.002	.000
310	.000	.001	.010	.071	.233	.352	.244	.078	.011	.001
350	.000	.000	.002	.021	.116	.293	.339	.181	.044	.005
390	.000	.000	.000	.004	.041	.173	.336	.299	.122	.025
430	.000	.000	.000	.001	.010	.073	.237	.352	.241	.087

JACKSON FARM

-------------------------------------Mean Value of Returns----------------------------------

	$20	40	60	80	100	120	140	160	180	200
$20	.228	.159	.180	.167	.126	.078	.039	.016	.005	.002
40	.180	.143	.176	.176	.144	.095	.051	.023	.008	.003
60	.138	.126	.167	.180	.159	.114	.066	.032	.012	.005
80	.104	.108	.154	.180	.171	.132	.083	.143	.018	.008
100	.076	.089	.138	.174	.178	.149	.101	.056	.025	.013
120	.055	.072	.120	.163	.181	.163	.120	.072	.035	.020
140	.038	.056	.101	.149	.178	.174	.138	.089	.047	.030
160	.026	.043	.083	.132	.171	.180	.154	.108	.061	.043
180	.017	.031	.066	.114	.159	.180	.167	.126	.077	.061
200	.011	.023	.051	.095	.144	.176	.176	.144	.095	.085

Field-Specific Stochastic Returns

The final step in preparing the returns data for analysis was to adapt the field-specific average returns to the ten returns states developed for each farm. The mean returns from the time series data did not match the mean per acre returns based on field-specific averages (Tables 6.3 and 6.4), probably due to differences in the time periods involved as well as to the differences in productivity implied by focusing on only a few soil types in each county.

To create ten levels of returns for each field, the field's per acre returns were subtracted from the time-series whole-farm mean per acre returns to generate an offset value. The offset value was then subtracted from each of the ten values associated with each returns level presented in Table 6.5. Because the ineligible acres will always be farmed, those fields can be treated as one large block of land for the purposes of calculating per acre returns. Individual totals are maintained for each of the eligible fields because those fields can be either farmed or enrolled. Returns were calculated for the ineligible blocks and for each eligible field; calculated values are presented in Table 6.5.

The field-specific per acre returns can be weighted by field size to estimate average per acre returns for each returns level for the entire farm. These estimates are before-tax figures, of course, since after-tax returns cannot be calculated before whole-farm returns are known. They are used to label levels of returns in tables, even though they are not after-tax values, because they convey more information about the returns states than simply numbering the levels as 1 through 10. The estimates appear immediately under the heading, "Mean Value of Last Year's Returns."

Once ten levels of returns were defined for each field on a representative farm, these figures could be used with the transition probabilities in order to calculate the discounted present value of future returns:

Table 6.5. Ten Levels of Per Acre Returns to Farming for Each Field

Field	Returns State									

CHAMPAIGN FARM

			---Mean Value of Last Year's Returns---							
	$47	87	127	167	207	247	287	327	367	407
1-3	50	90	130	170	210	250	290	330	370	410
4	37	77	117	157	197	237	277	317	357	397
5	-3	37	77	117	157	197	357	277	317	357
6	-35	5	45	85	125	165	205	245	285	325

JACKSON FARM

			---Mean Value of Last Year's Returns---							
	$22	42	62	82	102	122	142	162	182	202
1-8	17	37	57	77	97	117	137	157	177	197
9	69	89	109	129	149	169	189	209	220	249
10	38	58	78	98	118	138	158	178	198	218
11	-21	-1	19	39	59	79	99	119	139	159

$$PV_t = R_{i,t} + \beta \sum p_{ij} R_{j,t+1}$$

where

t	=	year index (stage);
i	=	current returns state; and
j	=	future returns state.

In this model, the producer is concerned with whole-farm, after-tax returns, so the above calculation is performed at the farm level after the individual field returns have been aggregated.

Tax Calculations

In order to choose between the farming and CRP alternatives for each eligible field, the whole-farm, after-tax alternatives must be compared for each field. Thus, returns by field must be aggregated to their before-tax total and the tax calculations completed in order to make the comparison. For this analysis, taxes were calculated for a married couple with two children filing jointly. The taxes paid include self-employment tax, federal income tax, and Illinois state income tax. No capital gains or losses, investment tax credits, or corporate taxes are included in the calculations.[14]

Results

For each farm, CRP enrollment decisions are examined for the following baseline scenario: the producer does not participate in commodity programs, uses a 5 percent discount rate to calculate present value over a 15 year planning horizon, and makes CRP enrollment decisions based on the county bid caps. Once the baseline is established, higher and lower discount rates are examined to determine their influence on CRP enrollment decisions, and county bid caps are varied to evaluate their impact. The analysis concludes by comparing enrollment decisions based on before-tax returns with those based on after-tax returns to see whether a whole-farm, after-tax model is actually required to investigate policy alternatives, or whether a field-specific model would suffice.

The Baseline Scenario

Table 6.6 presents the enrollment decisions made on each farm under the baseline scenario during the five-year CRP enrollment period. Since each farm has three eligible fields, the table shows a decision for each field for each level of returns. Thus, each row of the table contains ten sets of decisions, with each set consisting of a decision for each of the eligible fields. The decisions are displayed as 0 (out of the CRP, being farmed) and I (in the CRP, for one to five years). For example, the decision set for the Champaign farm in 1986 when last year's mean returns were $47 per acre is 00I: farm Fields 4 and 5, and enroll field 6.

It is readily apparent from Table 6.6 that only the poorest fields on these representative Illinois farms are likely to be considered viable candidates for the CRP. On the Champaign farm, Field 6 does not enter unless the previous year's mean returns to farming were $167 per acre or less. Field 5 enters in1988 when mean returns to farming were $47 per acre or less, and Field 4 is always farmed.

Table 6.6 Land-Use Decisions for the Three CRP-Eligible Fields on the Champaign and Jackson Farms

Year	Optimal Land Use Decision[a]									

Champaign Farm
Fields 4,5,6

			----Mean Value of Last Year's Returns----							
	$47	87	127	167	207	247	287	327	367	407
1990	OII	00I	00I	00I	000	000	000	000	000	000
1989	OII	00I	00I	00I	000	000	000	000	000	000
1988	OII	00I	00I	00I	000	000	000	000	000	000
1987	00I	00I	00I	00I	000	000	000	000	000	000
1986	00I	00I	00I	00I	000	000	000	000	000	000

Jackson Farm
Fields 9,10,11

			----Mean Value of Last Year's Returns----							
	$22	42	62	82	102	122	142	162	182	202
1990	00I	00I	00I	00I	00I	00I	00I	000	000	000
1989	00I	00I	00I	00I	00I	00I	00I	00I	000	000
1988	00I	00I	00I	00I	00I	00I	00I	00I	000	000
1987	00I	00I	00I	00I	00I	00I	00I	00I	000	000
1986	00I	00I	00I	00I	00I	00I	00I	00I	000	000

[a]I = field in CRP
0 = field being farmed

Field 11 on the Jackson farm is more likely to be enrolled in the CRP than any other field because its soil quality is so poor that it is in compliance with conservation standards only when wheat and hay are incorporated into the crop rotation. Requiring this rotation drastically lowers mean returns for this field. Field 11 enters the CRP in all but the highest returns states. Fields 9 and 10 are less erosive, so erosion can be controlled to acceptable levels with a corn/soybean rotation which employs conservation tillage. Fields 9 and 10 are always farmed under these circumstances.

Variations of the Baseline Scenario

Repeating the analysis with discount rates of 2 percent and 10 percent showed that the CRP enrollment decision was not highly sensitive to the magnitude of the discount rate. With only three eligible fields, changes are harder to identify than they would be with a larger number of eligible fields representing a greater continuum of productivity and erodibility. However, the analyses did show[15] that as the discount rate increased, land was more likely to be enrolled when returns to farming were low and less likely to be enrolled when they were high.

Surveys of producers who did not participate in the CRP indicate that many consider the rental rates inadequate compensation for withdrawing their fields from production. To test the impact of higher rental rates on the enrollment decision, the analysis was repeated using the approved rates of $115 and $70 per acre for the Champaign and Jackson farms, respectively. Hypothetical bid caps of $125 per acre in Champaign and $100 per acre in Jackson were also tested.

A bid cap of $115 in Champaign County drew Field 5 into the CRP when last year's returns were $127 per acre, and Field 6 when the returns were below $247. At a cap of $125, those fields entered at slightly higher levels of returns, and even Field 4 was drawn in at the lowest level. Field 11 on the Jackson farm entered at all levels of returns with a bid cap of $70, and Field 10 entered at the two lowest levels when the bid cap reached $100. Field 9 was never enrolled in the CRP.[16]

Reluctance to enroll Fields 9 and 10 on the Jackson farm can be attributed to the fact that those fields represent that farm's most productive land. The contrast in enrollment decisions for the Champaign and Jackson farms is particularly interesting because in terms of productivity and erodibility characteristics, Field 5 roughly parallels Field 9, and Field 6 roughly parallels Field 10. This contrast strikingly illustrates how the government's policy of paying different rental rates for land of similar quality makes rental rate rather than erodibility the primary factor in the land composition of the CRP.

Before-Tax and After-Tax Decisions

The analyses were repeated using before-tax returns as a basis for the enrollment decision to see whether the decisions would vary from those based on after-tax returns. One would expect the decisions to differ when before-tax returns to farming or CRP enrollment served to push the producer into a higher tax bracket and actually reduced his after-tax returns relative to the other alternative.

Some of the scenarios analyzed gave identical decision rules in both cases, and in other cases the decision rules differed.[17] Overall, results were fairly consistent using the two sets of returns. Table 6.7 presents the enrollment decisions for the baseline scenario using before-tax returns rather than after-tax returns, which were used in making the decisions presented in Table 6.6. Using before-tax returns, Field 5 does not enter the CRP in 1988 as it does with after-tax returns. Similarly, Field 11 does not enter in 1989 and 1990 using before-tax returns. Otherwise, the enrollment decisions for the baseline scenarios are identical. The CRP enrollment alternative is slightly less attractive than farming when before-tax returns constitute the basis for making the decision.

In order to estimate the maximum financial loss associated with using before-tax rather than after-tax returns to make enrollment decisions, results from the most discrepant case were further evaluated. Returns from the after-tax decision rule were compared with returns using the before-tax decision rule in the after-tax model. In the context of this problem, the before-tax decision rule is by definition suboptimal. The greatest discrepancy in discounted net value of returns over the 15 year planning horizon was $31,789, which represents a difference of almost 6 percent between the two decision rules. The maximum expected value of the difference between the two decision rules was determined by multiplying the unconditional probability associated with each return state by the dollar difference between the before- and after-tax decisions (Howard 1960). The maximum expected cost of using the before-tax model to make CRP enrollment decisions for the Jackson farm was calculated to be $2,479 or $165 per year over a 15 year planning horizon.

The fact that most of the before- and after-tax decisions differ only slightly can be attributed in part to the Tax Reform Act of 1987. The new tax laws create a much more linear tax function with fewer discontinuities than previously prevailed. In this problem, the producer's entire income derives from his farm, and his taxes were free of carryover from previous years.

Table 6.7   Land-Use Decisions for the Three CRP-Eligible Fields on the Champaign and Jackson Farms Based on Before-Tax Calculations

Year	Optimal Land Use Decision[a]

Champaign Farm
Fields 4,5,6

----------------------------Mean Value of Last Year's Returns---------------

	$47	87	127	167	207	247	287	327	367	407
1990	0II	00I	00I	00I	000	000	000	000	000	000
1989	0II	00I	00I	00I	000	000	000	000	000	000
1988	00I	00I	00I	00I	000	000	000	000	000	000
1987	00I	00I	00I	00I	000	000	000	000	000	000
1986	00I	00I	00I	00I	000	000	000	000	000	000

Jackson Farm
Fields 9,10,11

--------------------------Mean Value of Last Year's Returns-------------------

	$22	42	62	82	102	122	142	162	182	202
1990	00I	00I	00I	00I	00I	00I	00I	000	000	000
1989	00I	00I	00I	00I	00I	00I	00I	000	000	000
1988	00I	00I	00I	00I	00I	00I	00I	00I	000	000
1987	00I	00I	00I	00I	00I	00I	00I	00I	000	000
1986	00I	00I	00I	00I	00I	00I	00I	00I	000	000

[a]I = field in CRP
0 = field being farmed

When the producer's income includes depreciation, capital gains, or off-farm income, tax considerations may become more important. However, as income from other sources increases,[18] the tax impact of the enrollment decision for a single field decreases in relative importance.

Thus, the empirical and practical rationale for using a whole-farm model to analyze CRP enrollment decisions is less than compelling. A field-specific model would accommodate more state variables, stochastic and/or deterministic, and thus incorporate more factors into the decision process. The biggest advantage of using a field-specific model would be having the option to analyze as many fields on a farm as desired. The number of eligible fields that could be evaluated would not be limited, as it is with the whole-farm model in its present configuration. A field-specific model could be expected to yield optimal decisions for the majority of fields, but some decisions would be suboptimal from an after-tax standpoint.

Conclusions

CRP enrollment decisions made in a stochastic, dynamic framework differ from those which prevail in a static problem with certain returns. When returns are stochastic, they are likely to change over time--low returns tend to rise in the future, and high returns tend to fall. When farming and CRP alternatives are considered over a multiyear planning horizon, these probable trends in net returns affect the attractiveness of the CRP alternative relative to the static problem. When current returns are low, a producer viewing the decision myopically is more likely to enroll fields than a producer viewing the problem over a longer horizon. Similarly, when returns are high, the producer with the static viewpoint is less likely to enroll.

The probability transition matrices calculated for each of the representative farms constitute another advantage of using a stochastic, dynamic model, and represent a potential contribution to policy making. In the last section of this paper, the unconditional probabilities were used to calculate the maximum expected cost of using the before-tax model to make CRP enrollment decisions. Transition probabilities can also be used to calculate the probability of any particular field entering the CRP, given any particular starting level of net returns to farming.

Using a whole-farm model as a basis for decision making is probably not necessary in this application, however. A field-specific model would permit more variability in prices, yields, and costs, and the majority of decisions resulting from such a model should be optimal both before and after taxes. If the federal government extends the Conservation Reserve Program in the 1990 Farm Bill beyond the currently legislated limits, this could be a fruitful avenue for future research.

Notes

1.   Use of the male pronoun is not intended to imply that all producers are men; many women earn a living quite capably in farming. But because the English language lacks a suitable gender-free pronoun, and because men do predominate in the profession, the "producer" in this paper is referred to as male.

2.   Strictly speaking, both R_t and C are uncertain due to the unknown path of future inflation, but the uncertainty associated with inflation will not be considered in this analysis.

3.   Theoretically, a farmer can withdraw his land by paying a penalty. In practice, the penalty is so severe that withdrawal is not a viable option. Because the government's money must be repaid with interest, commodity prices would have to rise extremely high, particularly during the end of the contract period. According to Dr. Richard Farnsworth, agricultural economist at the University of

Illinois, a few farmers have defaulted on their contracts but none have bought them back from the government.

4. A framework for analyzing the enrollment decisions of a commodity program participant is presented in Allard (1989). Because the stochastic returns variable did not adapt well to the inclusion of program requirements, the case of the participant farmer was omitted from this paper.

5. The random variables do not enter the objective function in a linear manner (Taylor 1983).

6. U.S. Department of the Treasury, 1987.

7. The number of eligible fields was limited to three for this analysis because a program modeling that number could be run on a personal computer.

8. Based on soil productivity indices as described in Fehrenbacher et al.

9. For this analysis, CRP eligibility is based on potential erosion (RKLS/T) as computed by the State of Illinois Soil Conservation Service.

10. The number of eligible fields was restricted to three because of the modeling considerations discussed in the previous section.

11. See Allard (1989), Table 4-1, for field-specific cost of production data.

12. When discounting is used, averaging will give a different result than annually alternating crops. However, since there is no a priori reason to begin the model with one crop rather than the other, averaging is used in this analysis as a compromise solution.

13. Tables of these data are presented in Appendix E of Allard (1989).

14. A complete explanation of the federal taxes used in this model can be found in the *Farmer's Tax Guide* published by the US Department of the Treasury, Internal Revenue Service.

15. See Allard (1989) for tables and further details of the results.

16. Tables of these results are presented in Allard (1989).

17. A series of comparative tables are available in Allard (1989).

18. By 1985 off-farm income averaged 56 percent of total farm income (USDA 1985).

References

Allard, Celia A. *Optimal Participation in the Conservation Reserve Program: Land Enrollment Decisions in a Stochastic, Dynamic Framework.* Unpublished dissertation, University of Illinois at Urbana-Champaign, 1989.

Burt, Oscar R. "Farm Level Economics of Soil Conservation in the Palouse Area of the Northwest." *American Journal of Agricultural Economics* 63(1980):83-92.

Cooperative Extension Service, College of Agriculture, University of Illinois at Urbana-Champaign. *Illinois Agronomy Handbook*, Circular 1266, 1987-1988.

Fehrenbacher, J.B., R.A. Pope, I.J. Jansen, J.D. Alexander, and B.W. Ray. *Soil Productivity in Illinois.* Cooperative Extension Service, College of Agriculture, University of Illinois at Urbana-Champaign. Circular 1156, 1978.

Gallagher, P. and R.C. Green. *A Cropland Use Model: Theory and Suggestions for Estimated Planted Acreage Response.* USDA, Economic Research Service, Natural Economics Division, November, 1984.

Glaser, Lewrene K. *Provisions of the Food Security Act of 1985.* National Economics Division, Economic Research Service, USDA. Agricultural Information Bulletin 498, 1986.

Illinois Department of Agriculture, Division of Natural Resources. "T by 2000: A State Plan for Meeting "T" or Tolerance Soil Losses in Illinois by the Year 2000." Springfield, IL, 1985a.

_____. "Soil and Water Conservation District Administrative Guidelines for the Illinois Conservation District Administrative Guidelines for the Illinois Conservation Practices Program (CPP) and the Illinois Watershed Land Treatment Program (WLTP) for Cost-Sharing Erosion Control." Springfield, IL, 1985b.

Kennedy, John O.S. "Applications of Dynamic Programming to Agriculture, Forestry, and Fisheries: Review and Prognosis." *Review of Marketing and Agricultural Economics* 49(1981):141-73.

Lee, M.T., A.S. Narayanan, and E.R. Swanson. *Economic Analysis of Erosion and Sedimentation, Upper Embarras River Basin.* Department of Agricultural Economics, Agricultural Experiment Station, University of Illinois at Urbana-Champaign. AEER #136, 1975.

Narayanan, A.S., M.T. Lee, and E.R. Swanson. *Economic Analysis of Erosion and Sedimentation, Lake Glendale Watershed.* Department of Agricultural Economics, Agricultural Experiment Station, University of Illinois at Urbana-Champaign. AERR #131, 1974.

Seitz, Wesley D. and Earl R. Swanson. "Economics of Soil Conservation from the Producer's Perspective." *American Journal of Agricultural Economics* 63(1980):1084-88.

Swanson, E.R. and D.E. MacCallum. "Income Effects of Rainfall Erosion Control." *Journal of Soil and Water Conservation* 24(1969):56-9.

Taylor, C. Robert. "A Description of AGSIM, an Econometric-Simulation Model of Regional Crop and National Livestock Production in the United States." College of Agriculture and Alabama Agricultural Experiment Station, Auburn University. Staff Paper ES89-1. January, 1989.

_____. "A Flexible Method for Empirically Estimating Probability Functions." *Western Journal of Agricultural Economics* 9(1984):66-76.

_____. "Certainty Equivalence for Determination of Optimal Fertilizer Application Rates with Carry-over." *Western Journal of Agricultural Economics* 8(1983):64-7.

U.S. Congress, Public Law 99-198, 1985.

USDA, Agricultural and Rural Economy Division. *Economic Indicators of the Farm Sector: National Financial Summary.* ECIFS 5-2, 1985.

_____. *Economic Indicators of the Farm Sector: Costs of Production, 1979-1986.*

USDA, Soil Conservation Service in cooperation with Illinois Agricultural Experiment Station. *Soil Survey of Champaign County, Illinois.* Illinois Agricultural Experiment Station Soil Report 114, March, 1982.

_____. *Soil Survey of Jackson County, Illinois.* Illinois Agricultural Experiment Station Soil Report 106, February, 1979.

USDA, Soil Conservation Service. "Cropland Resource Management Systems." IL-CPA-11, December, 1987.

_____. *Field Office Technical Guide*, Section V, April, 1986.

7

A Profit Comparison of Range Culling Decisions in the Southwest

Russell Tronstad and Russell Gum

Pregnancy testing of range cattle is usually done to determine if cows should be culled from the herd. That is, the traditional standard rule of thumb when pregnancy testing is to sell cows that are open and keep cows that are pregnant since feeding a cow for almost another year in order to receive a live calf is generally equated with losing money. Trapp, Innes and Carman conclude that other factors such as the age of the cow and market prices will also influence an optimal culling decision rule in a dynamic framework. However, these studies did not consider the pregnancy state of the cow in determining optimal culling decisions. Consequently, the objective of this analysis is to solve for dynamically optimal culling decisions that consider the pregnancy state of the cow, age of cow, and stochastic nature of current market conditions (i.e., replacement price, culling price, and calf price), and compare the expected discounted present value of returns from these optimal decisions to the traditional rule of culling whenever a cow is open.

Optimal culling decisions are generated under the assumption of a multi-period horizon where the producer is presumed to maximize expected wealth (i.e., risk neutrality).[1] Two commonly utilized frameworks which incorporate the elements of time and uncertainty in decision problems are Stochastic Dynamic Programming (SDP) and optimal control theory. SDP is selected over optimal control theory for this analysis since an explicit solution is often not achieved for many problems when using optimal control theory (Burt) and this problem is exacerbated when stochastic variants are present (Whittle).

The SDP recursive equation for determining optimal culling decisions is presented in the next section. Joint calculation of calf and replacement price transition probabilities is discussed in the third section, while the fourth section discusses other critical input features and values of the culling model. Optimal culling decisions and the expected present value of optimal culling compared to culling when open are presented in the fifth section. Finally, a section of concluding comments discusses some implications for researchers and producers in making optimal culling decisions.

Stochastic Dynamic Programming Model

Because cows that are open can biologically be pregnant and detected as pregnant within six months, a bi-annual model is utilized. Also, since the analysis is applied to Southwest range conditions, feeding and calving requirements are not nearly as seasonally demanding in the Southwest as for many Northern states with harsh winters, where an annual model might be more appropriate. In the Southwest, a two season breeding program allows a rancher to carry an open cow for six months before rebreeding instead of the year required using an annual breeding season. The months of November and May are utilized as points in time for making the culling decision since these months tend to match up well with the pregnancy testing time associated with the current time framework utilized by many ranchers for late winter and late summer calving, respectively.

When specifying any model, an evaluation needs to be made between the trade off between model complexity and the value of information added to a model that is more complex. In considering the stochastic nature of replacement, culling, and calf prices in the SDP model formulation, culling prices were hypothesized to be adequately captured by current replacement and calf prices in the tradeoff between model complexity and more precise information. That is, cull prices are hypothesized to be primarily driven by a demand to be maintained in the breeding herd (captured by replacement price information) or slaughtered for meat (adequately captured by current calf prices). The following equation supports this hypothesis;[2]

(1) $SL_t = -2.502 + .0462 \cdot REP_t + .3538 \cdot CALF_t \quad \bar{R}^2 = .944$
$\quad \quad (-.7499) \quad (6.756) \quad \quad (4.7279) \quad \quad D\text{-}W = 1.618$

where SL_t is the monthly average price of slaughter cows, cutter 1-2 ($/cwt.) at Cottonwood, California, REP_t is the contemporaneous replace price of cows for Red Bluff and Shasta, California auction prices combined (*Livestock and Meat Prices and Receipts at Certain California and Western Area Markets*), $CALF_t$ is the combined steer and heifer calf price for California ($/cwt., *Agricultural Prices*), $\bar{R}^2$ is the adjusted coefficient of determination and D-W is the Durbin-Watson statistic. All prices are deflated by the consumer price index (1967=100) and prices run from 1971 to 1988. Observations were missing on replacement prices for May in 1976, 1977, and 1983 so that 33 observations were available for estimating equation (1).

Additional lags and seasonal factors were found to be insignificant in equation (1). Since model complexity is reduced by a factor of at least seven in making slaughter prices deterministic, the trade off between model complexity and a more precise decision rule appears to weigh heavily in support of a deterministic slaughter relationship. This is especially true in that slaughter value is more of a residual factor than replacement or calf price values in the culling decision.

The objective function for this Southwest rancher is to maximize the expected present value of profit over a bi-annual (November-May) T-period planning horizon subject to the state variables of the cow's age, replacement price, calf price, and state (i.e., pregnancy/calf/condition) of the cow. Formulation of this problem as a SDP model results in the following recursive equation:

$$(2) \quad V_t(CAGE_t, REP_t, CALF_t, PCC_t = \underset{K-R_t}{MAX} \; E[\, Ret_t(CAGE_t, REP_t, CALF_t,$$

$$PCC_t + b \cdot V_{t+1}(CAGE_{t+1}, REP_{t+1}, CALF_{t+1}, PCC_{t+1})\,]$$

Subject to:

(3) $CAGE_{t+1} = CAGE_t + 1$ if $K-R_t = K$ [Age relationship]
(i.e., cow is Kept in the herd)

(4) $CAGE_t \leq 11$ years in age   [Maximum age of cow allowed for breeding]

(5) $K-R_t = R$ (i.e., replace) if
$PCC_t \geq 6$ or the cow is not   [Breeding fitness constraint]
suitable for breeding in t+1.

(6) $SL_t = -2.502 + .0462 \cdot REP_t$ [Deterministic slaughter
$+ .3538 \cdot CALF_t$ price relationship]

(7) $Ret_t = Calfweight(CAGE_t) \cdot CALF_t$
$- Cost(CAGE_t)$ [Return function]
if $K-R_t = K$; else
$Ret_t = Salvage(CAGE_t, REP_t, CALF_t, PCC_t) - Cost(CAGE_t) - REP_t$

(8) $(REP_{t+1}, CALF_{t+1}) = f_1(REP_t, CALF_t)$ [Joint stochastic Markovian transitions associated with replacement and calf prices]

(9) $PCC_{t+1} = f_2(PCC_t)$ [Stochastic Markovian cow state relationship]

where t is a bi-annual time index; $V_t(\bullet)$ is the maximum expected value of returns from period t through the terminal period T given the initial state; E is the expectation operator; Ret_t is the current return function; b is the discount factor (1/1.03 - approximately 6% annual discount factor); $K-R_t$ is the decision variable of whether to Keep (K) or Replace (R) a cow; $CAGE_t$ is the cows current age (1.5 to 11 years old); PCC_t is the current state of the cow (i.e., pregnancy/calf/condition); Calfweight($\bullet$) is the weight of the calf (100 lbs.) as a function of $CAGE_t$; Cost($\bullet$) is the cost of maintaining a cow for six months, dependent on $CAGE_t$; Salvage($\bullet$) is the salvage value of liquidating a cow and any calf from the herd, thus salvage depends on $CAGE_t$, REP_t, $CALF_t$, and PCC_t; $f_i(\bullet)$'s (i=1,2) are stochastic Markovian relationships; and all other variables are as described earlier.

Equation (2) simply says that the optimal value function in period t (i.e., $V_t(\bullet)$) is the maximum of current returns (i.e., $Ret_t(\bullet)$) plus the discounted optimal value function in period t+1 (see Howard). Equation (3) accounts for the future consequences of our current decision in that the cow will be six months older in the following period (t+1) if the cow is not culled in t. Equation (4) constrains the cow's age from exceeding 11 years. Equation (5) indicates that the cow must be replaced if the current state of the cow is unfit for breeding. The deterministic slaughter price relationship described above is imposed by equation (6). Current returns from keeping or replacing a cow are captured in equation (7) while stochastic Markovian processes are described in equations (8) and (9).

Joint Calculation of Replacement and Calf Price Transition Probabilities

To determine Markovian transition probabilities for replacement (REP_t) and calf prices ($CALF_t$), these prices were estimated as a function of own lagged prices, cross prices, and a dummy variable for the month of May. Own and cross lagged prices are hypothesized to capture current market conditions while a dummy variable is included for the month of May to capture any bi-annual (seasonal) influences. In estimating REP_t and $CALF_t$, variables were omitted if they were not statistically significant.

Table 7.1 Markovian Conditional Price Transition Probabilities for May and November in Going from a Replacement Price of $640/head and Calf Price of $85/cwt. in t to Other Replacement and Calf Prices in Period t+1, for May and November

Calf Price	MAY Replacement Price ($/Hd. in t+1)				
($/cwt. in t+1)	$ 420	$ 530	$ 640	$ 750	$ 860
$ 65	0.0000	0.0000	0.0000	0.0000	0.0000
$ 75	0.0000	0.3191	0.2711	0.0570	0.0000
$ 85	0.0000	0.0136	0.1159	0.1336	0.0000
$ 95	0.0000	0.0002	0.0133	0.0760	0.0000
$105	0.0000	0.0000	0.0000	0.0000	0.0000
Calf Price	NOVEMBER Replacement Price ($/Hd. in t+1)				
($/cwt. in t+1)	$ 420	$ 530	$ 640	$ 750	$ 860
$ 65	0.0000	0.0000	0.0000	0.0000	0.0000
$ 75	0.0000	0.1168	0.0142	0.0003	0.0000
$ 85	0.0000	0.1597	0.1327	0.0150	0.0000
$ 95	0.0000	0.0564	0.2504	0.2543	0.0000
$105	0.0000	0.0000	0.0000	0.0000	0.0000

Also, it was detected that the error structure between REP_t and $CALF_t$ was significantly correlated. Consequently, the following two equations were determined to be appropriate and were estimated by seemingly unrelated regression:[3]

(10) $\quad CALF_t = .8951 \cdot CALF_{t-1} + 15.512 \cdot DMAY_t + e_{1t} \qquad \bar{R}^2 = .693$
$\qquad\qquad (36.507) \qquad\quad (7.058) \qquad\qquad\qquad$ D-W = 2.473

(11) $\quad REP_t \;\; = 132.49 + 1.1816 \cdot REP_{t-1} \qquad\qquad \bar{R}^2 = .782$
$\qquad\qquad (3.575) \quad (13.524) \qquad\qquad\qquad\qquad$ D-W = 2.031

$\qquad\qquad - .4098 \cdot REP_{t-2} + e_{2t}$
$\qquad\qquad (-4.7583)$

where $DMAY_t$ is a dummy variable for the month of May, e_{it} (i=1,2) is a normally distributed error term with covariance $\sigma_{12} \neq 0$ (e.g., see Judge, et al.), and all other variables and statistics are as defined earlier. The second order Markov process of REP_t was reduced to a first order Markov process, as described in Burt and Taylor, utilizing the reproductive property of a normal distribution. Normality was not rejected for equation (10) or (11) at a five percent significance level utilizing either a goodness of fit test for normality or the Jarque-Bera asymptotic LM normality test.

Romberg's double numerical integration algorithm in Gerald and Wheatley was modified to account for the dependence between e_{1t} and e_{2t}, in calculating price transition probabilities from equations (10) and (11). Table 7.1 gives an example of the joint Markovian transition probabilities calculated for the months of May and November.

Other Critical Input Features and Values

As described in Table 7.2, the cow's age is related to the probability that the cow will need to be culled (i.e., culling necessary from diseases, lameness, and other factors), die, or remain suitable for breeding purposes. Also, expected calf percentages of pregnant cows are related to the cow's age.

These are some of the critical input features suggested by expert opinion of an animal science specialist[4] that are utilized in determining the Markovian transition probabilities in going from one cow state to another cow state, as shown in Table 7.3. As delineated in Table 7.3, a cows current state can take on any of the following 8 state conditions of; 1) pregnant with a sale calf at side, 2) pregnant with no calf at side, 3) open with a sale calf at side, 4) open with no calf at side, 5) open with a newborn calf at side, 6) unfit to breed with a sale calf at side, 7) unfit to breed with no calf at side, and 8) dead with no calf at side. Table 7.3 also indicates the probability associated in going from one of these eight state conditions in period t to another state condition in period t+1. It is noted that the probability associated with a cow becoming pregnant is greater if it doesn't have a newborn calf at its side (i.e., .84 versus .79), simply due to the increased nutrition requirements of the calf.

Other critical values included in the model are: sale calf weight of 400, 425, and 450 lbs. for cows that are 1.5-2.0, 2.5-3.0, and 3.5-11 years old, respectively; bi-annual feed costs of $125.00 and $100.00 for cows that are 1.5-2.0, and 2.5-11 years old, respectively; culling weight of 800 lbs. for all cows; and a terminal or salvage value of $80.00 assigned to a newborn calf. It is assumed that all replacements bought are pregnant and 1.5 years old. Replacement prices are discretized into the five price increments of $420, $530, $640, $750, and $860 per head, while calf prices are discretized into the five price increments of $65, $75, $85, $95, and $105 per hundredweight. These prices roughly correspond to the mean price plus one and two standard deviations above and below the mean. Mid-points between price states are utilzed for integrating the probability of going from one REP$_t$-CALF$_t$ state to all other possible REP$_{t+1}$-CALF$_{t+1}$ states.

Replacement and calf price transitional probabilities are multiplied by the state transition probabilities in Table 7.3 to get the combined joint probability in going from one REP$_t$-CALF$_t$-PCC$_t$ state in t to all other possible REP$_{t+1}$-CALF$_{t+1}$-PCC$_{t+1}$ states in period t+1.

Optimal Culling Decisions Versus Culling When Open

This section presents optimal converged culling decisions from the previously presented SDP model. Optimal culling decisions are only given for the cow states of open and no calf at side (state 4) or open and sale calf at side (state 3) since all of the other states had essentially a constant decision rule to always keep the cow if it had a newborn calf at its side and always keep the cow if it was pregnant, providing that the cow wasn't over 9 years old and replacement prices were above $420 per head. This result emphasizes the importance of considering the pregnancy state of the cow in conjunction with market price considerations, unlike what previous recent studies have done (e.g., Trapp, Innes and Carman). Because the decision rule was the same for cow states 3 and 4 above, only one decision rule is given for the Spring (May) and Fall (November) pregnancy test time periods in Table 7.4.

Optimal culling decisions are given in Table 7.4 for four different cow ages (i.e., 2, 6, 8.5, and 10 years), and all possible replacement and calf price states. Although, the probability of being at a very high replacement price and low calf price or vice versa, is not very high, Table 7.4 gives all possible price combinations to illustrate the importance of price in optimal culling decisions. For both May and November, results indicate that one should only replace an open 2 year old if calf prices are above $85/cwt. when bred replacements can be purchased for $420/head or if calf prices are above $95/cwt. when the replacement price is $530/head. However, the decision to replace instead of keep an open cow increases quite rapidly as the cow approaches 8 years in age.

The decision to replace occurs less frequently for November than May, due to calf prices being estimated as seasonably higher for May than November (see equation 10). That is, cows which test open in November and are kept in the herd, should be tested as pregnant in May, calve in six months, and have a sale size calf at their side in May when calf prices are most seasonably favorable. Whereas, a replacement purchased in November would have a sale size calf at its side the following November when calf prices are not as seasonably favorable. At an age of 10 years, the decision rule is to always replace an open cow, except for November's highest replacement price and lowest calf price combination.

Table 7.4 Optimal Culling Decisions (K and R Denote Keep and Replace, Respectively) for May and November with Different Replacement Prices, Calf Prices, and Various Cow Ages, Given that the Cow is Not Pregnant and Suitable for Breeding

MAY

Cow's Age = 2 years Replacement Price ($/Head)						Cow's Age = 6 years Replacement Price ($/Head)				
Calf Price ($/cwt.)	$420	$530	$640	$750	$860	Calf Price ($/cwt.)	$420	$530	$640	$750	$860
$ 65	K	K	K	K	K	$ 65	R	K	K	K	K
$ 75	K	K	K	K	K	$ 75	R	R	K	K	K
$ 85	R	K	K	K	K	$ 85	R	R	K	K	K
$ 95	R	R	K	K	K	$ 95	R	R	R	K	K
$105	R	R	K	K	K	$105	R	R	R	R	K

Cow's Age = 8.5 years Replacement Price ($/Head)						Cow's Age = 10 years Replacement Price ($/Head)				
Calf Price ($/cwt.)	$420	$530	$640	$750	$860	Calf Price ($/cwt.)	$420	$530	$640	$750	$860
$ 65	R	R	R	K	K	$ 65	R	R	R	R	R
$ 75	R	R	R	R	K	$ 75	R	R	R	R	R
$ 85	R	R	R	R	R	$ 85	R	R	R	R	R
$ 95	R	R	R	R	R	$ 95	R	R	R	R	R
$105	R	R	R	R	R	$105	R	R	R	R	R

NOVEMBER

Cow's Age = 2 years Replacement Price ($/Head)						Cow's Age = 6 years Replacement Price ($/Head)				
Calf Price ($/cwt.)	$420	$530	$640	$750	$860	Calf Price ($/cwt.)	$420	$530	$640	$750	$860
$ 65	K	K	K	K	K	$ 65	R	K	K	K	K
$ 75	K	K	K	K	K	$ 75	R	K	K	K	K
$ 85	R	K	K	K	K	$ 85	R	R	K	K	K
$ 95	R	R	K	K	K	$ 95	R	R	R	K	K
$105	R	R	K	K	K	$105	R	R	R	K	K

Cow's Age = 8.5 years Replacement Price ($/Head)						Cow's Age = 10 years Replacement Price ($/Head)				
Calf Price ($/cwt.)	$420	$530	$640	$750	$860	Calf Price ($/cwt.)	$420	$530	$640	$750	$860
$ 65	R	R	R	K	K	$ 65	R	R	R	R	K
$ 75	R	R	R	R	K	$ 75	R	R	R	R	R
$ 85	R	R	R	R	K	$ 85	R	R	R	R	R
$ 95	R	R	R	R	K	$ 95	R	R	R	R	R
$105	R	R	R	R	R	$105	R	R	R	R	R

Table 7.5 gives the expected discounted present value of returns from following an optimal culling strategy for 15 years and a more traditional strategy of culling whenever a cow is open for 15 years. If one weights the values in Table 7.5 by the unconditional probabilities, determined non-parametrically from the 1971-1988 calf-replacement price series, results indicate that an optimal culling strategy averages a net present value of $345.86 more than a strategy of culling whenever the cow is open. Utilizing a 6% annual discount rate, a rancher could expect to increase the profitability on the current carrying capacity of his/her range or lease by $35.61/head per year by following optimal culling decisions generated from the SDP model presented over a more traditional strategy of culling whenever a cow is open. Also, as one would expect, the benefit of following the optimal strategy over the more traditional strategy is greatest (least) at high (low) replacement prices and low (high) calf prices, since these price combinations are most conducive to keeping (replacing) an open cow (see Table 7.4).

Conclusions

Contrary to tradition, cows were frequently kept when they were tested as open. Cows that tested pregnant were kept in the breeding

Table 7.5 Expected Discounted Present Value of Returns from Following an Optimal Culling Strategy or a Culling Strategy of Replacement Whenever the Cow is Open (values in brackets) over a 15 Year Planning Horizon for Different Initial Replacement and Calf Prices. Values are Given for the Initial State of a One and a Half Year Old Heifer that Is Pregnant

Calf Price ($/cwt.)	Replacement Price ($/Head)				
	$420	$530	$640	$750	$860
$ 65	53.69 [-252.39]	43.18 [-291.10]	19.69 [-376.90	0.71 [-446.90]	-7.32 [-476.81
$ 75	86.08 [-211.83]	74.99 [-250.71]	50.88 [-337.47]	31.46 [-408.76]	23.24 [-439.19]
$ 85	141.20 [-211.83]	128.12 [-176.13]	102.81 [-263.16]	82.90 [-334.85]	74.55 [-365.48]
$ 95	201.93 [-59.28]	184.79 [-98.02]	156.47 [-195.05]	135.76 [-256.85]	127.23 [-287.55]
$105	252.73 [-12.73]	219.46 [-51.37]	188.45 [-138.38]	166.99 [-210.20]	158.18 [-240.90]

herd, unless they were approaching the maximum age and replacement prices were low. This result indicates the importance of considering the pregnancy state of the cow when making optimal culling decisions. Consequently, contrary to recent culling models presented in the literature, researchers should consider the pregnancy state of the cow in deriving optimal replacement decisions.

The SDP model presented can be solved on a 16-bit microcomputer with 640K of random access memory. Therefore, ranchers could easily modify this program for their specific biological situation and determine optimal culling decisions that could be a viable management tool for increasing their profits. Results suggest that the rancher would be able to improve profitability by almost $36/head per year by utilizing an optimal culling decision as opposed to a more traditional strategy of culling whenever the cow is open.

Notes

1. In part, this analysis utilizes a risk neutral objective function, since the theoretical underpinnings of incorporating risk in a dynamic setting are not clearly sorted out (Kreps and Porteus, and Mossin), and many dynamic model decisions yield results that appear risk averse (e.g., Antle; Just; and Pope).
2. t-values are in parentheses.
3. t-values are in parentheses.
4. Dick Rice, personal communication.

References

Antle, J. M. "Incorporating Risk in Production Analysis." *American Journal of Agricultural Economics* 62(1982):89-97.

Burt, O. R. "Dynamic Programming: Has Its Day Arrived?" *Western Journal of Agricultural Economics* 7(1982):381-94.

Burt, O. R., and C. R. Taylor "Reduction of State Variable Dimensions in Stochastic Dynamic Optimization Models Which Use Time Series Data." *Western Journal of Agricultural Economics* 1990.

Howard, R. A. *Dynamic Programming and Markov Processes.* The Massachusetts Institute of Technology: John Wiley & Sons, Inc., New York, 1960.

Innes, R. and H. Carman "Tax Reform and Beef Cow Replacement Strategy" *Western Journal of Agricultural Economics* 13(1988):254-66.

Judge, G. G., R. Hill, W. E., Griffiths, H. Lutkepohl, C. Lee. *Introduction to the Theory and Practice of Econometrics*, pp. 321-325. New York: John Wiley & Sons, 1982.

Just, R. E. "Risk Aversion under Profit Maximization." *American Journal of Agricultural Economics* 57(1975):347-52.

Kreps, D. M., and E. L. Porteus, "Temporal Resolution of Uncertainty and Dynamic Choice Theory." *Econometrica* 46(1978):185-200.

Mossin, J., "A Note on Uncertainty and Preferences in a Temporal Context." *American Economic Review* 59(1969):172-4.

Pope, R. D., "Empirical Estimation and Use of Risk Preferences: An Appraisal of Estimation Methods that Use Actual Economic Decisions." *American Journal of Agricultural Economics* 64(1982):376-83.

Rice, D. Professor of Animal Science, University of Arizona. Personal communication.

Trapp, J. N. "Investment and Disinvestment Principles with Nonconstant Prices and Varying Firm Size Applied to Beef-Breeding Herds." *American Journal of Agricultural Economics* 68(1986):691-703.

Whittle, P. *Optimization Over Time: Dynamic Programming and Stochastic Control.* Vol. 1. New York: John Wiley & Sons, 1982.

8

Optimal Farm Program Participation and Base Acreage Adjustment Under Alternative Program Provisions

Patricia A. Duffy

Participation in the farm program can drastically alter farmers' income, costs, and crop-mix. In deciding whether to participate, producers must consider prevailing prices, acreage reduction rates, and limits on plantings of program crops imposed by participation. In exchange for direct payments and other program incentives, participants are limited to planting only a portion of their eligible "base" acreage in each program commodity. Under the 1981 farm bill, a producer's eligible base in a commodity was a simple two-year average of acreage planted and considered planted, where acreage "considered planted" is acreage idled because of the provisions of the program for that commodity. Under the 1985 farm bill, the base is calculated as a five-year average, with the stipulation that the base can not exceed the two year average. Under the 1985 farm bill, therefore, producers would be much more limited in their ability to increase their eligible acreage of a profitable program crop.

The objective of this study is to examine the optimal decision rules for crop mix and participation in the farm bill for a representative Alabama cotton and soybean farm under alternative base calculation provisions, two-year and five averages. The problem is formulated as a multi-stage dynamic programming problem. The paper builds upon previous work by Mims et al. by considering the stochastic nature of cotton and soybean prices and by extending the five-year planning horizon used in the Mims study to twenty years. The twenty-year horizon was chosen because it is sufficiently long that further time consideration will not influence the current decisions.

Review of Literature

As early as 1948, Heady recognized that government program provisions influence crop mix decisions, but it was not until 1972 that a mathematical model of a farm firm included government program provisions. This model, developed by Scott and Baker, was a quadratic programming model of a central Illinois cash grain farm. Because program participation at that time reduced expected income, their analysis focused on the risk-income trade-off of participation versus nonparticipation. Several subsequent studies (Persaud and Mapp; Kramer and Pope; Musser and Stamoulis) focused on the same issue.

Currently, farm program participation usually increases expected income, depending on the percentage of acreage in eligible bases. On a per acre basis, it is almost always profitable to participate in the farm program. Participation, however, precludes planting any program crops beyond their program limits, and, thus, on a whole-farm basis, participation may reduce profits if the initial bases are small and nonprogram crops unprofitable.

The crop-mix and program participation decisions for any given year are complicated by the dynamic nature of the program base, which is an historic average of acreage in the commodity. Thus, opting out of the program and planting increased acreage in certain program crops may result in reduced profits in that year but increased expected profits over a longer time horizon. The multi-year decision environment for program crops has been examined by Perry and by Mims et al. Perry used quadratic programming to analyze the participation/crop-mix decision for Texas crop farms and found that with a two year base producers are often willing to exit the program for one or two years to adjust planting patterns, but with a five year base no adjustments took place. In analyzing the affect of risk aversion, Perry noted that the greater the assumed degree of risk aversion, the less likely the producers were to adjust planting patterns, even with a two year base.

Mims et al. developed five-year mixed integer programming models of Alabama cotton farms to examine how the change from a two- to a five- year base and the enforcement of limited cross-compliance affected crop mix/participation decisions under the hypothesis of profit maximization. Results from this study indicated that even with a five-year base and limited cross-compliance, some producers would still opt out of the program for a short time to adjust their bases.

Methods

The representative farm developed for this study is a cotton and soybean farm in Northwest Alabama. The farm is similar in many respects to one of the farms developed by Mims et al. but has been simplified for the dynamic programming formulation. Per acre yields and costs were developed using records from the Alabama Farm Analysis Association and budgets from the Alabama Cooperative Extension Service. Cotton yields were assumed to be 726 lb./acre, while soybean yields were assumed to be 28 bu./acre. Variable Costs are $263.50 per acre for cotton, $82.35 for soybeans, and $5.00 for any land idled due to farm program participation. Nonland fixed costs are $59.67/acre.

The target price for cotton is fixed at $0.70 for every year of the planning horizon. This target price is lower than the 1990 target price of $.729 to reflect expected further declines in future years. Unlike the Mims study in which price variability was not considered, here, market prices are treated as stochastic variables that follow a first order Markov process. In any given year of the planning horizon, the lagged market prices for soybeans and cotton are known but this year's prices are not known at planting time. Conditional probabilities of receiving a particular market price in the current year were thus developed from a set of econometric estimates linking lagged to current prices.

In any given year, the expected (mean) prices will be:

(1) $\quad E(PCT_t) = 0.96 * PC_{t-1}^{0.81588}$

(2) $\quad E(PSB_t) = 2.00 * PS_{t-1}^{0.66706}$

where PC is market price of cotton, PS is the market price of soybeans, E is the expectation operator, t is a time subscript, and EXP is the natural exponent function. The two prices are assumed to have a bivariate normal distribution with a standard deviations of 0.17869 for cotton and 0.1512 for soybeans and a correlation coefficient of 0.35044. From this information, the joint probability of receiving particular ranges of prices can be calculated using double numerical integration.

In any given year, the producer must decide whether or not to participate in the farm program and how to allocate acreage between cotton and soybeans. If the producer decides to participate in the farm program, the one-period profit function will be:

(3) $\quad \Pi = (MAX(TP, PC_t) * YC - VCC) * APGC_t - SA_t * VCSA$

$\qquad + (PSB_t * YS - VCS) * AS_t - FC$

where TP is the target price of cotton, YC is the yield of cotton, VCC is per acre variable costs of cotton, APGC are planted acres of cotton in the program, SA is land idled due to program participation, VCSA is the variable cost of idling the land, YS is the yield of soybeans, AS is acres of soybeans, and FC is fixed cost.

The program acres are limited by the base acreage, calculated as an historic average. In this study, the required acreage diversion ranges from 0.10 to 0.28% based on an inverse relationship to lagged market price. The exact formula is:

(4) $SA = 0.28 - \{(PC - .45)/0.05\}*0.03$

The diverted acreage is thus 0.28 at the lowest assumed cotton price ($0.45), and 0.10 at the highest possible cotton price ($0.75).

For a farmer who does not participate in the farm program, the one period profit function is:

(5) $\Pi_t = (PC_t*YC - VCC)*AC_t - + (PS_t*YS - VCS)*AS_t - FC$

where AC is nonprogram cotton acreage.

Because of the stochastic nature of prices, profits for any given combination of program participation, crop-mix alternative, are unknown at the time of the decision. Hence, an expected profit is calculated using the numerically integrated density function.

Dynamic Programming

Dynamic programming is a mathematical optimization technique for solving multistage decision problems (Bellman, 1957). In the dynamic programming formulation, decisions undertaken in one period will affect future returns. The whole time period over which decisions are made is called the planning horizon. Each interval in the planning horizon is referred to as a stage, and, in every stage, the system can be represented by a group of state variables. The planning horizon for the cotton farm problem is twenty years with each year representing a stage. State variables are the lagged cotton and soybean prices and the five year planting history that determine the base acreage. Thus, there are seven state variables in the problem.

The objective function for the problem can be expressed as:

(6)   Maximize $\{\Sigma_t \beta^{t-1} E(\Pi_t(PC_{t-1}, PS_{t-1}, Base_t, U_t))\}$

where Π_t is annual profit, U_t is the vector of decision variables, E is an expectations operator, $Base_t$ is the beginning cotton base (historic average of cotton acreage), and $ß^{t-1}$ is a discount factor. In this study, a discount factor of 0.090 (10% interest rate) was used.

The objective function can be expressed in recursive form as:

(7) $\quad V_t (PC_{t-1}, PS_{t-1}, A_{t-2}, A_{t-3}, A_{t-4}, A_{t-5}) = MAX_U \{E(\pi_t (PC_{t-1}, PS_{t-1}, Base_t, U_t))$

$\quad\quad + ßV_{t+1} (PC_t, PS_t, A_t, A_{t-1}, A_{t-2}, A_{t-3}, A_{t-4}.)\}$

where:

(8) $\quad Base_{t+1} = \Sigma_{i=0}^{N-1} A_{t-i}$

where A is the acreage of cotton, whether in or out of the farm program. (Program acreage includes the acreage reduced owing to participation.)

For the cotton farm, a recursive relationship exists between lagged market prices and the probability distribution of this year's price, but this relationship, unlike the base calculation, is unaffected by the decision variables. The price state variables were discretized over their assumed probable ranges. The minimum price of cotton was assumed to be $0.45 and the maximum $0.75. Five intermediate prices were used: $0.50, $0.55, $0.60, $0.65, and $0.70. Soybean prices were assumed to range from $5.00 to $9.00, also with five evenly distributed intermediate values. Historic planted acreage in cotton was represented as a percentage of all acreage. Thus, 0 and 1 were the bounds. Intermediate values of 0.25, 0.50, and 0.75 were chosen for the discrete intervals. In the five-year base calculation, five past acreage state variables are needed. For the two-year base, only two are required.

The main program begins by looping over the acreage state variables. Base acreage is then calculated. Next, the program loops over the price states and finds the optimal combination of the two decision variables. The participation/nonparticipation choice is discrete and is easily represented by a 0-1 integer. The portion of acreage to plant in cotton, however, is continuous and must, like the state variable, be discretized. The range from 0 to 1 was therefore subdivided by increments of 0.05. Because the five-year base is calculated by dividing the historic acreage by 5, it was necessary that the decision variable have a five times finer grid than the state variables.

Because this year's decision variable on planted acreage is next year's state variable for the one year lag, interpolation is used to improve the accuracy of translating the decision variable into the blockier state. If the decision variable were 0.05, for example, it would lie in the 0 to 0.25

range of the state variables, but much closer to the lower end. Linear weights are developed and the recursive equation (8) is expressed as

(9) $V_t = MAX_U \{E(\pi_t( ) + \text{ß}\{wl^*V_{t+1}(A_t(\text{lower}))$
$\qquad + wh^*V_{t+1}(A_t(\text{higher}))\}$

where wl is the weight for the lower value of the state range and $A_t(\text{higher})$ the value of the state variable closest to the decision variable but higher than it.

Optimal Decision Rules

Figures 1 and 2 are graphical representations of the optimal decision rules for the cotton farm under the alternative base acreage specifications. In Figure 8.1, program and nonprogram cotton acreage for the five-year base are presented for five possible bases, 0, 25%, 50%, 75%, and 100%. In Figure 2, acreages for the same set of possible bases are presented for the two-year base calculation. For the five-year base, there are many possible routes by which the farm could arrive at any given base. Cursory investigation, however, revealed little or no effect on the optimal decision rule. Hence, the optimal decision rules are presented only for the case of identical history (i.e, all previous years have planting patterns equal to current base.)

For the 0 beginning base situation, the primary difference between the two-and five-year base calculation is the level of lagged soybean price required before all acreage is switched from nonprogram cotton to soybeans at low lagged cotton prices. For the five-year base, soybeans are planted on all of the acreage when their price is greater than or equal to $7.67 and cotton price is $0.45. At higher cotton prices, the entire acreage is planted in cotton regardless of the soybean price. For the two-year base, acreage is switched to soybeans only when the lagged soybean price rises to $8.33 and the lagged cotton price falls to $0.45. Because current nonprogram plantings in a given year are worth more in a future year under the two-year rather than the five-year base, it is not surprising that it takes higher soybean prices to induce the producer not to plant cotton.

With a 25% base and a five-year calculation, 100% nonprogram cotton is always chosen when the lagged cotton price is greater than or equal to $0.55. At lower prices, acreage will be split between program cotton and soybeans, depending on the price of soybeans. With a two-year base calculation, the pattern is identical, but in every case it takes a higher soybean price to induce acreage out of nonprogram cotton.

Even with a fifty percent base, the optimal decision under the five-year base calculation is all nonprogram cotton whenever the cotton price of $0.55 or higher. For soybean prices below $7.00, 100% nonprogram cotton is planted for cotton prices at or above $0.50. The response for the two-year average has more diversity across soybean prices, but is otherwise no dissimilar.

With a 75% base, producers move to 100% nonprogram cotton at a $0.55 cotton price and low soybean prices, or at a $0.60 cotton price with higher soybean prices. Results for the two-year base calculation are quite similar, except that the levels of soybean prices at which strategies change are higher. With a 100% base, soybeans are never chosen and the strategies are identical for the two-year and five-year base calculations. 100% program cotton is planted until the market price exceeds $0.60, then 100% nonprogram cotton is planted.

Conclusions

Results of this study are quite different from those in Mims et al. where large changes in strategy resulted from a change from a two-year to a five-year base calculation. In this study, strategies were often the same.

In this study, soybeans were not an attractive alternative to nonprogram cotton, but would be planted in combination with program cotton if soybean prices were sufficiently high and cotton prices sufficiently low that the combination was more desirable than 100% nonprogram cotton. The major differences in this study found between the two- and five- year base was that farmers would be less willing to plant soybeans, even at very low cotton prices. Under the two-year base, farmers move to 100% nonprogram cotton somewhat more readily than under the five-year base. This result may have occurred because, under the two-year calculation, current nonprogram plantings translate into much higher future bases.

It is not altogether surprising that results here should differ substantially from those of Mims et al. That study used a much higher target price ($0.794/lb) making the farm program much more attractive than it was here. Also, the Mims et al. study included several program crops besides cotton, providing far more alternatives than were used here. Finally, the Mims et al. study ignored price variability and examined strategies based on expected future prices.

Extensions of the dynamic programming framework to analyze cotton program decisions should include more alternative crops than soybeans and should also examine yield variability. Sensitivity to yield levels and variable costs should also be examined. Finally, a tax function should be included to see how progressive tax rates affect the decision variables.

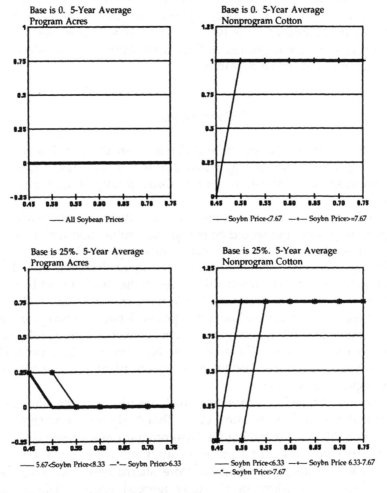

Figure 8.1 Cotton Average with Five-Year Base

Figure 8.1   continued.

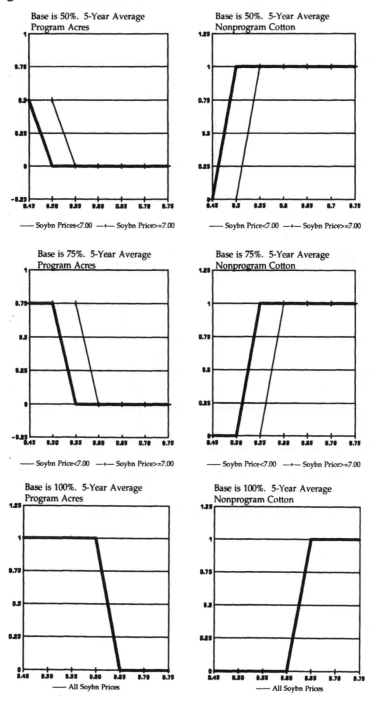

Base is 50%. 5-Year Average Program Acres

—— Soybn Prices<7.00 —+— Soybn Price>=7.00

Base is 50%. 5-Year Average Nonprogram Cotton

—— Soybn Price<7.00 —+— Soybn Price>=7.00

Base is 75%. 5-Year Average Program Acres

—— Soybn Price<7.00 —+— Soybn Price>=7.00

Base is 75%. 5-Year Average Nonprogram Cotton

—— Soybn Price<7.00 —+— Soybn Price>=7.00

Base is 100%. 5-Year Average Program Acres

—— All Soybn Prices

Base is 100%. 5-Year Average Nonprogram Cotton

—— All Soybn Prices

124

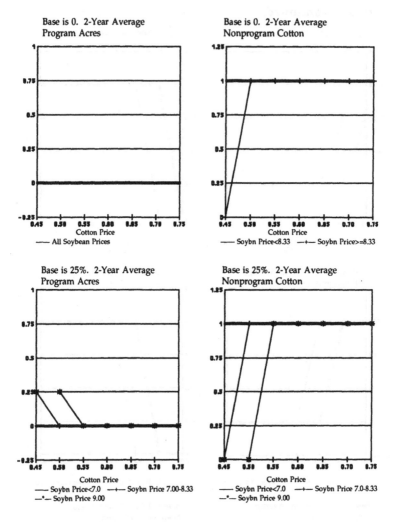

Figure 8.2 Cotton Average with Two-Year Base

Figure 8.2 continued.

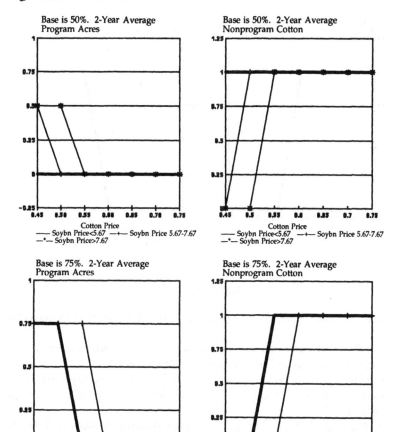

Base is 50%. 2-Year Average Program Acres

Cotton Price
—— Soybn Price<5.67 —+— Soybn Price 5.67-7.67
—*— Soybn Price>7.67

Base is 50%. 2-Year Average Nonprogram Cotton

Cotton Price
—— Soybn Price<5.67 —+— Soybn Price 5.67-7.67
—*— Soybn Price>7.67

Base is 75%. 2-Year Average Program Acres

Cotton Price
—— Soybn Price<7.67 —+— Soybn Price>=7.67

Base is 75%. 2-Year Average Nonprogram Cotton

Cotton Price
—— Soybn Price<7.67 —+— Soybn Price>=7.67

Figure 8.2 continued.

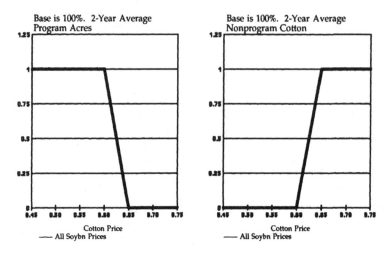

References

Bellman, R. *Dynamic Programming*. Princeton, NJ: The Princeton University Press. 1957.

Heady, E. O. "The Economics of Rotations with Farm and Production Policy Applications." *Journal of Farm Economics* 30(1948):645-64.

Kramer, R. A. and R. P. Pope. "Participation in Farm Commodity Programs: A Stochastic Dominance Analysis." *American Journal of Agricultural Economics* 63(1981):119-28.

Mims, A. M., P. A Duffy, G. Young. "Effects of Alternative Acreage Restriction Provisions on Alabama Cotton Farms." Journal Paper number 892052P of the Alabama Agricultural Experiment Station. April, 1989.

Musser, W. N. and K. G. Stamoulis. "Evaluating the Food and Agriculture Act of 1977 with Firm Quadratic Risk Programming." *American Journal of Agricultural Economics* 63(1981):448-56.

Perry, G. M. "Towards a Holistic Approach to the Cropping Mix Decision." Unpublished Ph.D. dissertation, Texas A&M University, August, 1985.

Persaud, T. and H. P. Mapp, Jr. "Analysis of Alternative Production and Marketing Strategies in Southwestern Oklahoma: A MOTAD Approach." *Risk Analysis in Agriculture: Research and Educational Development*. Department of Agricultural Economics, AE-4492, University of Illinois, June, 1980.

Scott, J. T., Jr. and C. B. Baker. "A Practical Way to Select an Optimum Farm Plan Under Risk." *American Journal of Agricultural Economics* 54(1972):657-60.

9

Stochastic Dynamic Programming Models: An Application to Agricultural Investment

Frank S. Novak and Gary D. Schnitkey

The nature of farm firms has special implications for the choice of risk management strategies. Most farms are small and have a noncorporate ownership structure which prevents spreading of risks among many individuals. Benefits derived from specialization in production and resource limitations restrict opportunities for enterprise diversification. This has spurred an interest in finding alternative methods of reducing the variability of cash flows and introducing greater diversity into the asset and liability structure of farm firms (Young and Barry). One method which has received limited attention is investment in financial assets. By allocating resources outside of the farm, cash flows may be stabilized without the inherent losses in productive efficiency which often accompany diversification into other agricultural enterprises. Young and Barry approached the problem of holding financial assets as part of a long term portfolio. Their results suggest that investment in financial assets may stabilize long term cash flows.

To date, very few studies have explicitly considered the stochastic, dynamic nature of investment returns in models of firm growth and investment (Schnitkey and Taylor; Larson, Stauber and Burt). Theoretical studies (Robison and Barry; Robison and Brake) have outlined the limitations of conventional portfolio theory as a farm planning and decision making tool. Since conventional portfolio analysis is static, it does not result in an operational investment strategy and ignores some key determinants of financial performance. Portfolio models fail to incorporate the implications of asset indivisibility, the liquidity characteristics of assets, tax impacts, and the costs of altering investment portfolios. Farm managers are concerned with the distinction between cash returns and asset appreciation and how these factors influence the firm's operating structure and growth. In a dynamic world, decisions made in one time period affect decisions in later periods by altering financial structure and the nature of productive assets. The dynamics of asset accumulation are thus a major point of interest as well. The analysis of investment decisions should incorporate these dynamic interrelationships in a multiperiod model.

The objective of this paper is to address the issue of dynamic investment problems in agriculture. Specifically, the potential effects of stock investment outside of an agricultural enterprise on a firm's financial structure are analyzed. The study also considers the influence of financial structure and returns to the agricultural enterprise on optimal investment decision rules. This is accomplished by specifying and numerically solving a stochastic dynamic programming (SDP) model for an Illinois hog finishing operation.

The paper also addresses some of the inherent problems associated with firm level financial models. Specifically, the issues of modelling bankruptcy and dealing with multiple stochastic state variables are considered. The calculation and use of conditional probability methods which have been adapted to the dynamic programming framework are also discussed and illustrated.

The Dynamic Programming Model

A monthly stochastic dynamic programming model was specified and solved to determine the optimal stock investment decision rule for an Illinois hog finishing operation. It was assumed that the decision maker could invest his funds into either stocks (S) or other financial instruments (OF) such as a money market fund. The stock investment could be easily achieved through a mutual fund. Returns to stock investment were calculated as changes in the price of the fund from period to period plus any dividend income. The model contained three stochastic state variables: hog returns (HR),defined as revenue minus variable costs, stock prices (PS), and return on other financial instruments (ROF), and two deterministic state variables: holdings of other financial instruments (OF), and stock holdings (S). Investments in other financial instruments represented an asset when held in positive amounts or a liability when negative (i.e., operating credit). The decision in each period was the level of stocks to buy or sell (DS).

The stock investment model was formulated as a terminal wealth maximization problem rather than the standard present value maximization. Denoting the terminal year as T, terminal wealth can be written as a function of the state variables:

$$V_T(HR_T, PS_T, ROF_T, OF_T, S_T) = (PS_T \bullet S_T) + OF_T + FarmAssets_T$$
$$- TermDebt_T$$

where $V_T(\bullet)$ is the recursive objective function for year T and FarmAssets were those assets devoted to the production of finished hogs and TermDebt is the level of long term debt. This function leads to the following general recursive equation:

$$V_{t-1}(HR_{t-1}, PS_{t-1}, ROF_{t-1}, OF_{t-1}, S_{t-1}) = \underset{DS_t}{Max}\ E[V_t(HR_t, PS_t, ROF_t, OF_t, S_t)]$$

where $E(\bullet)$ is the expectations operator and $V_t(\bullet)$ is the value of wealth assuming that optimal decisions are made.

This maximization is subject to the following state transition equations:

$$HR_t = f_1(HR_t)$$

$$PS_T = f_2(PS_{t-1}ROF_{t-1})$$

$$ROF_t = f_3(ROF_{t-1})$$

$$S_t = S_{t-1} + DS_t$$

$$OF_t = OF_{t-1} - With - DS_t \bullet PS_t\ DIV_t + HogRet_t$$

where the terms in the state transition equations are:

$$DS_t = \{200, 0, -200\}$$

Borrowing/Lending Differential (BLD) = 3%
If $OF_t \le 0$ then $ROF_t - ROF_t + BLD$
HogReturn$_t$ (HogRet$_t$ = HR$_t$ $\bullet$ MHP - FC $\bullet$ MHP

Withdrawals(With) = MCW + Payment on Term Debt
$0 \le S_t \le 2000$
$100 \le PS_t \le 310$
$-20 \le RH_t \le 40$
$-350000 \le OF_t \le 350000$
$.06 \le ROF_t\ .16$

Monthly Hog Production(MHP) = 750
Monthly Consumption Withdrawal(MCW) = 2,000
Fixed Cost Per Hog(FC) = 5.00
Stock Dividend(DIV$_t$) = .083•S$_{t-1}$
Farm Assets$_1$ = 450,000
Bankruptcy ≡ {Wealth ≤ 0}
Beginning TermDebt = 100,000

Rather than discounting returns, as in a present value maximization, returns in the terminal wealth maximization problem are compounded. Within the stock investment model, compounding is achieved through the other financial instruments holding variable (OF).

Estimation of Transition Probabilities

Numerical solution of the investment model required state transition probabilities which were derived from estimated state transition equations. This section describes the data and estimation procedures for the hog return (HR), stock price (PS) and interest rate (ROF) state transition equations.

Monthly hog returns were based on budgets reported in the *Livestock Meat Situation and Outlook Report* published by USDA. Data from the Illinois Farm Business Farm Management Association (FBFM) were used to adjust the return series to reflect Illinois costs of production as closely as possible.

Monthly stock prices (S&P 500 index) were collected from the Standard and Poor's Statistical Reporting Service and dividend data were based on information provided by Ibbotson and Associates. Short term interest rate data were from the *Economic Report of the President*. All series covered the period from the beginning of 1974 to the third quarter of 1987.

Modelling of multiple stochastic state variables provides a special problem in that transition relationships originate from multivariate stochastic processes. Thus, a state variable's transition relationship may include not only its own lagged variables but other lagged state variables as well.

The nature of the economic variables in this model provided an additional problem. An index was used to represent stock prices. The index showed a continual upward trend which was the result of economic growth. Likewise, interest rates, expressed in nominal terms, have also trended upwards over the sample period. In terms of time series analysis, the existence of trends results in non-stationary data.

Stationarity is required to ensure that the estimated transition probabilities are derived from a process which is time invariant. Stationarity was achieved by differencing the stock price and interest rate series once.

Tentative dynamic interrelationships between hog returns, stock prices and interest rates were originally identified through time series techniques (Granger and Newbold). Sample autocorrelations and cross-correlations suggested that hog returns were not correlated with either interest rates or stock prices. Hog returns showed evidence of lower order autoregressive structure. Autocorrelations and partial autocorrelations for stock price and interest rate suggested that these variables were interrelated and that lower level autoregressive models would adequately capture their Markovian relationships.

Hog Return Transition Relationship

Examination of autocorrelations and partial autocorrelations suggested that the hog return relationship could be modelled with a second order autoregressive process, AR(2). A goal of reducing the number of state variables prompted the estimation of a first order process (AR(1)) as well as the AR(2) model.

Estimation of the AR(1) model produced the following equation (t-statistics in parentheses):

$$HR_t = 1.895 + .811HR_{t-1} \quad R^2=.684 \sigma_e=8.003$$
$$\quad\quad (2.58) \quad (18.70)$$

where σ_e is the standard error of the estimate. This formulation resulted in autocorrelated errors indicating that an AR(1) model did not adequately capture the series' time dependent nature.

Estimation of the AR(2) model resulted in the equation:

$$HR_t = 2.430 + 1.177HR_{t-1} - .439HR_{t-2} \quad R^2=.736 \sigma_e=7.239$$
$$\quad\quad (3.60) \quad (16.54) \quad\quad (-6.29)$$

which showed no sign of autocorrelation and yielded normally distributed errors as judged by the Jarque-Bera test statistic.

Based upon these results, and analyses of higher order models, the AR(2) model was judged to adequately describe the series' Markovian nature. To reduce the dimension of the DP model only one hog return variable was included. The reduction was accomplished using Burt and Taylor's method of reducing the order of an autoregressive process. This procedure resulted in the following form:

$$HR_t = 1.688 + .8177HR_{t-1} \quad \sigma=8.058$$

From this equation, transition probabilities were estimated using a hyperbolic tangent method (Taylor).

Table 9.1 shows the resulting transition probability matrix for hog returns as well as the limiting distribution which is approached within one year of any beginning state level. This table illustrates the extreme variability of returns in hog feeding enterprises. This variability provided some of the rationale for the choice of a monthly rather than an annual model.[1]

Interest and Stock Price Transition Relationships

Autocorrelations, partial-correlations and cross-correlations were examined to identify autoregressive relationships within the first differenced stock price and interest rate series. These plots suggested a first order autoregressive structure for each variable and across variables. Estimation of the AR(1) model for stock price resulted in the following parameters (t-statistics in parentheses; all variables in log form):

$$PS_t - PS_{t-1} = .124(PS_{t-1} - PS_{t-2}) - .126I_{t-1} + .118I_{t-2}$$
$$(3.13) \qquad\qquad (-3.63) \quad (3.27)$$
$$\sigma_e=.035$$

which reduced to:
$$PS_t = 1.124PS_{t-1} - .124PS_{t-2} - .126I_{t-1} + .118I_{t-2}$$

Estimation of the AR(1) model for first differenced interest rates resulted in the following parameters:
$$I_t - I_{t-1} = .328(I_{t-1} - I_{t-2})\sigma_e=.0636$$
$$(4.95)$$

which reduced to:
$$I_t = 1.328I_{t-1} - .328I_{t-2}$$

Table 9.1 Transition Probabilities for Hog Returns

	HR_{t+1}				
HR_t	-20	-5	+10	+25	+40
-20	.6063	.3760	.0176	.0001	.0000
-5	.1064	.6217	.2644	.0075	.0000
+10	.0032	.1778	.6462	.1699	.0029
+25	.0000	.0082	.2744	.6167	.1007
+40	.0000	.0001	.0190	.3869	.5940
Limit	.0719	.2544	.3786	.2342	.0608

Residuals from both of the above equations were normally distributed as judged by the Jarque-Bera statistic and were independent across equations based on cross-correlations.

As was the case with the hog return transitions, a reduction in the number of state variables was preferred to lower the dimension of the DP model. Burt and Taylor's method for reducing the order of interdependent autoregressive equations was employed to produce the following equations:

$$S_t = S_{t-1} - .0064 I_{t-1} \sigma_e = .0381$$
$$I_t = I_{t-1} \qquad \sigma_e = .0717$$
$$\sigma_{s1} = -.0035 \qquad \Rightarrow \rho_{s1} = -.1267$$

From these equations transition probabilities for stock prices and interest rates were estimated using a numerical integration routine (Gerald and Wheatley).

Optimal Stock Investment Decision Rule

The optimal stock investment decision rule was derived using a value-iteration dynamic programming algorithm. Numerical solution required specification of discrete state and decision variable levels. Four hog return intervals ranging from -$20 to $40 produced state levels of -20, 0, 20, and 40 dollars respectively. Stock prices covered 15 intervals ranging from 100 to 310 and stock holdings ranged from 0 to 2,000 units in increments of 200. Financial instrument holdings covered the range -$350,000 to $350,000 in $70,000 increments and return on financial instruments ranged from 6 percent to 16 percent in two percent increments. This formulation resulted in 43,560 states. The stock purchase decision was allowed to take on values of -200 (sell), 0, or 200 (buy).[2]

The optimal investment rule was obtained by backward recursion beginning at the final year of the planning horizon. Linear interpolation of the objective function was used to increase the convergence rate and reduce biases resulting from discretizing the state variables. Interpolation was used on the financial holdings variable because the ending values for this variable did not necessarily match the state interval midpoints. Optimal decisions were found for all state intervals except those combinations which defined technical bankruptcy (i.e. negative wealth). In the case of bankruptcy, the farming operation was presumed to be liquidated. Optimal decision rules were generated until the optimal decision rules converged, which occurred by month six of year three. Thus, the converged decision rule was applicable to all periods up to the thirty months before the end of the planning horizon. For example, if the planning horizon is ten years long, the converged decision rule would be applicable from year one through to month six of year seven. The large size of the optimal decision rule prevents a complete description within this paper. A graphical presentation of a portion of the decision rule follows in figures 9.1 and 9.2.

Figure 9.1 presents the optimal decision rule when hog returns are at the $20 level and stock holdings are 400 units. Panels A through C illustrate the effects of changing stock price levels for the complete range of interest rates and financial holdings.

For given levels of stock price and financial holdings, the graphs illustrate the dampening effects of higher interest rates on the desirability of stock purchases. For example, at a stock price of $115 (Panel A) and financial holdings of $0, stock purchases occur up to an interest rate of 12 percent. Over a range of 12 to 14 percent, existing stocks are held, and stocks are sold at rates above 14 percent.

This interest rate effect exists at all stock price levels although the absolute values of interest rates at which decisions change vary with stock price. For fixed levels of stock price and interest rates, the graphs illustrate the effects of higher financial holdings levels on stock purchases. As financial holdings increase, the firm has more funds to purchase stocks. At a 12 percent interest rate and stock price of 115, for example, financial holdings of less than -$70,000 are associated with stock sales. The range from -$70,000 to $70,000 are associated with a decision not to purchase or sell stocks, and levels above $70,000 are associated with stock purchases. As was the case with interest rates, this wealth effect is consistent across stock price levels although the breakpoints differ for each stock price.

Figure 9.2 illustrates the effects of three different hog return levels on stock purchases. At constant stock prices of $190 and stock holdings of 400 units, panels A through C illustrate the effects of three different hog return levels across the range of interest rates and financial holdings. Note that the interest rate and financial holdings effects discussed above occur for different hog return levels. The effect of changes in hog returns is seen in the positioning of buy and sell breakpoints. For example, at an interest rate of ten percent,a hog return of -$20 (panel A) implies almost no purchases at lower levels of financial holdings. A return level of $0 (panel B) implies purchases at lower levels of financial holdings up to an interest rate of 12 percent. A $20 return also allows purchases up to an interest rate of 12 percent at most levels of financial holdings and infers no stock sales until 12 percent even at the lowest level of financial holdings.

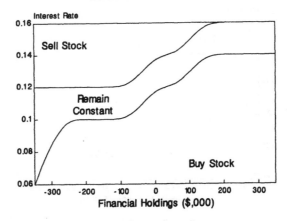

Panel A. Stock Price = $115

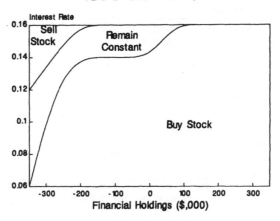

Panel B. Stock Price = $160.

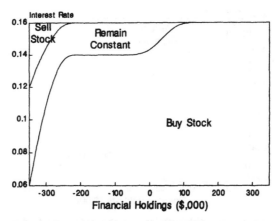

Panel C. Stock Price = $220.

Figure 9.1 Portions of the Optimal Decision Rule for Hog Returns of $20 and Stock Holdings of 400

136

Panel A. Hog Return = -$20

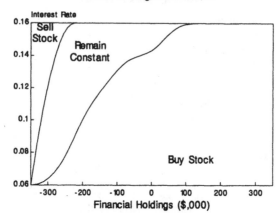

Panel B. Hog Return = $0

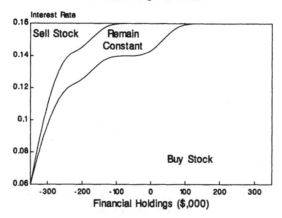

Panel C. Hog Return = $20

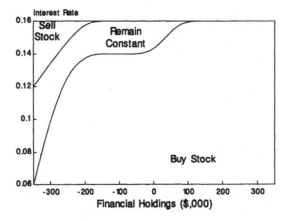

**Figure 9.2 Portions of the Optimal Decision Rule for
Stock Holdings of 400 and Stock Price of 190**

Calculation of Conditional Probabilities

Discrete conditional probability methods were used to determine ex ante distributional forecasts of financial holdings and stock holdings. The general approach for calculating conditional probabilities requires constructing an N dimensioned state probability vector denoted as Π_t, where N is the number of state intervals and t equals the number of the stage. The i'th element of this vector $(\pi_t(i))$ gives the probability of being in the i'th state at time t. To satisfy the basic properties of probabilities, each element must be between zero and one and the sum of all elements must equal one. The state probability vector for the initial state Π_0 contains one element that is one while the rest are zero, indicating that the beginning state is known with certainty.

The movements of the system between points in time are given by an N by N transition matrix (P). The rows of P correspond to states at the current point in time while the columns correspond to states at the next point in time. The elements of the i'th row give the probability of moving from state i to any of the N possible states at the end of the period.

Multiplication of the initial state probability vector by the transition matrix yields the state probability vector for the next time period:

$$\Pi_L = \Pi_{L-1}P$$

The i'th element of the Π_1 vector gives the probability of being in state i in period one conditional on the probabilities in the Π_0 vector. In general, the transition during time period L is given by:

$$\begin{aligned} \Pi_L &= \Pi_{L-1}P \\ &= \Pi_{L-2}PP \\ &= \Pi_0 P^L \end{aligned}$$

where P^L indicates that the P matrix is postmultiplied L times. The vector gives the probabilities of being in each of the states at time L. These probabilities represent an ex ante forecast given the initial state.

The investment model's state probability vector contains five dimensions with each element represented as:

$$\pi_t(hr, ps, rof, of, s)$$

where hr, ps, rof, of, and s are state interval indices for hog returns, stock price, interest rate, financial holdings and stock holdings respectively.

The same state variable discretation was used in solving the conditional probabilities as was used in solving the dynamic programming model. In addition, a bankruptcy state was added. This state accumulated the probability associated with state intervals resulting in bankruptcy. Thus, each state probability vector contained 43,561 elements (4 hog return states x 15 stock price states x 6 interest rate states x 11 financial holdings states x 11 stock holdings level states + 1 bankruptcy state). Each element of the state probability vector representing a solvent state was referenced as:

$$\pi_t(hr,ps,rof,of,s)$$

where hr, ps, rof, of, and s were state interval indices for the hog return, stock price, return on other financial assets, financial holdings and stock holdings level respectively. The final element represented bankruptcy.

The transition matrix was constructed using the state transition equations and the optimal stock investment decision rule. This matrix was a square matrix of dimension 43,561. The 43,561st row represented bankruptcy. The remaining rows of the transition matrix were decomposed as follows:

$$[HR] \otimes [PSR] \otimes [OFS_{hr,ps,rof}]$$

where [HR] was the hog returns state transition matrix, [PSR] was the stock price-interest rate state transition matrix, and $[OFS_{hr,ps,rof}]$ were matrices giving other financial holdings and stock holdings. This partitioning was possible because the [HR] matrix depended only on the hog returns state transition equation and the [PSR] matrix depended only on the stock price and interest rate state transition equations. The [HR] matrix was a square matrix of dimension 4. The [PSR] matrix was a square matrix of dimension 90 (15 stock price states x 6 interest rate states). The $[OFS_{hr,ps,rof}]$ matrices were square matrices with dimension 121 (11 financial holding states x 11 stock holding states). To calculate these matrices, the hog return, stock price, and return on other financial asset states had to be known. Thus, there were 360 matrices, with one matrix corresponding to each hog return, stock price, and return on other financial asset combination. These matrices were calculated using all state transition equations and the optimal decision rule.

Within the computer program which calculated conditional probabilities, the [HR] and [PSR] matrices were calculated once and stored in RAM. Elements within the $[OFS_{hr,ps,rof}]$ were calculated as needed. By repeating the above process, conditional probabilities were found for a five-year horizon.

Resulting state probability vectors represented a joint probability density function, conditional on the beginning state variable levels, presuming that the optimal decision rule was followed. Standard discrete probability techniques (see, for example, Hogg and Craig) were used to find a state variable's marginal distribution. This required finding the probability of being in each of the state variables' intervals. For example, the probability of occupying the of_i financial holdings state interval equaled:

$$Pr(of_i) = \sum_{hr} \sum_{ps} \sum_{for} \sum_{s} \pi_t \ (hr, ps, rof, of_{i}, s)$$

Repeating the above operation for all financial holdings state intervals resulted in the marginal distribution. Marginal distributions were found for financial holdings and stock holdings levels.

Application of conditional probability theory (Hogg and Craig) to the marginal distributions developed above yields important information on the linkages between the state variables. The probability of being in the of_i financial holdings state interval given a particular stock holding, s_i is calculated as:

$$Pr(of_i \mid s_i) = \frac{\sum_{hr} \sum_{ps} \sum_{for} \pi_t(hr, ps, rof, of_{i}, s_i)}{Pr(of_i)}$$

Completing this calculation for all financial holdings states gives the financial holdings distribution conditional on the stock holdings state s_i.

The effects of bankruptcy were investigated using this method. Financial holdings and stock holdings distributions conditional on being solvent were calculated. For example, the conditional probability of being in the of_i state equaled:

$$Pr(of_i | solvent) \quad = Pr(of_i) \ / \ (1 - Pr(bankrupt))$$

where Pr(bankrupt) equals the probability in the bankruptcy state. Repeating the above operation for all state intervals resulted in the conditional distribution. A similar process was conducted for the stock holdings interval.

Also, expected wealth levels were calculated using the state probability vector. This was accomplished by summing the result of each state interval's probability times each state interval's wealth.

Conditional Probability Results

Conditional probabilities were calculated for three different initial financial holdings levels: -$70,000, $0, and $70,000. The initial values for the remaining state variables were the same for the above three conditional probability calculations. These state variables were a $10 hog return, a 130 stock price, a 10 percent interest rate, and a 0 unit stock holding level.

Panel A of Figure 9.3 shows marginal financial holdings distributions at the end of years 1, 3, and 5 for an initial financial holdings level of $0. At the end of year 1, the majority of the probability is in the negative financial holdings region, indicating that debt has been accumulated. Most of this debt results from stock purchases, as illustrated in panel B. This panel shows the probability associated with each stock holdings level. As can be seen, the majority of the probability is associated with stock holding levels greater than 1,600 units.

Over time, probabilities associated with financial holdings and stock holdings become more evenly distributed across the respective states. This is illustrated by the general flattening of both the financial holdings and stock holdings distributions in years 3 and 5 (Figure 9.3). More evenly distributed probability results because of wider possible ranges of stock prices and returns on other financial instruments. As adverse stock prices result, stock holdings will be reduced, resulting in higher financial holdings.

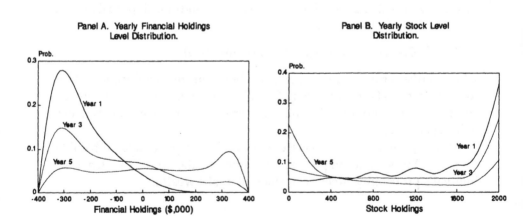

**Figure 9.3 Yearly Conditional Distributions $0
Beginning Financial Holdings**

A correlation exists between stock holdings and other financial holdings. This is illustrated in Figure 9.4, which shows financial holdings distributions conditional on stock holdings levels of 0, 600, and 1,800. As can be seen, higher stock holdings are associated with lower financial holdings. Higher stock holdings require debt purchases, yielding the resulting skewed distributions. Note also that the 0 stock holding level has considerable probability associated with financial holdings levels above $200,000. This high probability, along with the .23 marginal probability of having 0 stock holdings, suggest that stock price and interest rate combinations exist in which stock is not a wise investment. This result is supported by the decision rule presented earlier.

The bankruptcy probability in year 5 is .3343 for the initial financial holdings level of $0. As initial financial holdings increase, the probability of bankruptcy decreases. For example, initial financial holdings of -$70,000, $0, and $70,000 result in year 5 bankruptcy probabilities of .4564, .3097, and .2441, respectively. Initial financial holdings levels also have large impacts on wealth in year 5. Expected wealth levels of $292,467, $380,210, and $467,016 respectively result from initial financial holdings levels of -$70,000, $0, and $70,000.

The interaction between initial financial holdings, bankruptcy and ending financial and stock holdings distributions was of great interest. The results above indicated that beginning financial holdings affect both bankruptcy and ending stock and financial holdings distributions. In order to isolate the effect of bankruptcy on these distributions, we

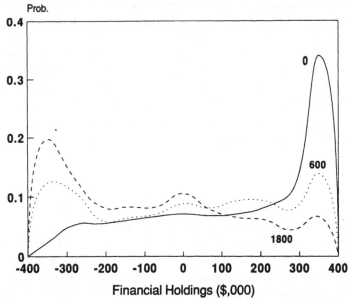

Figure 9.4 Financial Holdings Given Differing Stock Levels, Year 5

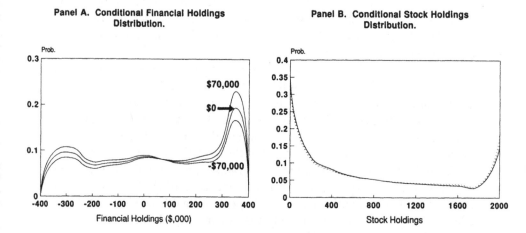

Panel A. Conditional Financial Holdings Distribution.

Panel B. Conditional Stock Holdings Distribution.

**Figure 9.5 Conditional Distributions for Stock
and Financial Holdings, Year 5**

calculated these distributions conditional on solvency. The resulting distributions (Figure 9.5) indicate that initial financial holdings have little impact on the distributions once bankruptcy is taken into account

Higher initial wealth results in more probability being in the higher financial levels (panel A). This difference in probability, however, is relatively small when compared to the $140,000 difference in initial financial holdings. Stock holdings are much more closely distributed (panel B). The financial holdings and stock holdings distributions tend to converge towards some particular distribution, given that the firm is solvent and that the optimal stock investment rule is followed.

Conclusions

The first objective of this paper was to address the issue of off-farm investment in a dynamic framework, incorporating the effects of farm financial structure and market conditions. A second objective was to illustrate some of the important issues and techniques which arise when developing and solving stochastic dynamic firm level decision models.

Methods were presented for estimating transition probabilities for interrelated stochastic state variables and reducing model dimensions. Conditional probability methods which enhance the information available from stochastic dynamic programming models were discussed and illustrated.

The dynamic programming model of a hog finishing operation identified the effects of interest rate levels, stock prices and financial structure on optimal stock purchases. Higher interest rates were shown to dampen stock purchases over all ranges of financial holdings and stock prices. The absolute levels for buy/sell breakpoints were dependent upon both financial holdings and stock prices. The model also identified the effects of different hog return levels on investment decisions. Greater profits in hog production implied greater stock purchases for given levels of interest rates or financial holdings.

Conditional probability methods were used to project future financial structure and investment holdings for different beginning states. A correlation between stock holdings and financial holdings was identified and illustrated. Greater stock holdings implied the use of larger amounts of short term debt. Initial financial holdings were shown to affect the probability of bankruptcy and expected terminal wealth but had little influence on distributions of financial and stock holdings.

It is worth noting that static investment models would only provide information on the mix of investments given some static objective function. The dynamic investment model provides an operational investment strategy as well as providing ex ante forecasts of future financial structure and investment mix when the optimal decision rule is followed. The extra information obtained from the dynamic model should be of interest to decision makers.

Notes

1.   Quarterly and annual models were investigated for all three stochastic state variables but no Markovian structure could be identified.

2.   The model was also solved using a finer grid with 6 hog return states and 11 interest rate states for a total of 127,776 states. The results were essentially the same as for this smaller version.

References

Burt, O.R. and C.R. Taylor. "Reduction of State Variable Dimension in Stochastic Dynamic Optimization Models Which Use Time Series Data." Working Paper No. 88-2. Dept. of Agricultural Economics, University of California, Davis, 1988.

Gerald, C.F. and P.O. Wheatley. *Applied Numerical Analysis*. Third Ed. Addison-Wesley Publishing Co. Don Mills, Ontario, 1984.

Granger, C.W.J. and P. Newbold. *Forecasting Economic Time Series*. Second Ed. Academic Press, Inc. New York, 1986.

Hogg, R.V. and A.T. Craig. *Introduction to Mathematical Statistics*. Fourth Ed. MacMillan Publishing Co., Inc., New York, 1978.

Howard, R.A. *Dynamic Programming and Markov Processes*. John Wiley and M.I.T. Press. New York, 1960.

Ibbotson Associates. *Stocks, Bonds, Bills, and Inflation. 1988 Yearbook*. Chicago, Illinois, 1988.

Larson, D.K., M.S. Stauber and O.R. Burt. "Economic Analysis of Farm Firm Growth in Northcentral Montana." Research Report No. 62. Montana Agricultural Experiment Station. Montana State University, Bozeman, Montana, 1974.

Robison, L.J. and J.R. Brake. "Application of Portfolio Theory to Farmer and Lender Behavior." *American Journal of Agricultural Economics* 61(1979):158-64.

Robison, L.J. and P.J. Barry. "Portfolio Theory and Asset Indivisibility: Implications for Analysis of Risk Management." *North Central Journal of Agricultural Economics* 2,1(1980):41-6.

Schnitkey, G.D. and C.R. Taylor. "Optimal Farmland Investment Decisions: A Dynamic Programming Analysis." Staff Paper No. 87 E-384. Dept. Of Agricultural Economics, University of Illinois at Urbana-Champaign, 1987.

Standard and Poor's Corporation. *Statistical Service, 1988*.

Taylor, C.R. "A Flexible Method for Empirically Estimating Probability Functions." *Western Journal of Agricultural Economics* 9(1984):66-76.

United States Dept. of Agriculture. *Livestock and Meat Situation and Outlook Report*. Selected Issues.

U.S. President. *Economic Report of the President*. Selected Issues.

Young, R.P. and P.J. Barry. "Holding Financial Assets as a Risk Response: A Portfolio Analysis of Illinois." *North Central Journal of Agricultural Economics* 9,1(1987):77-84.

10

Modeling Dynamics and Risk Using Discrete Stochastic Programming: A Farm Capital Structure Application

Allen M. Featherstone, Timothy G. Baker,
and Paul V. Preckel

The choice of capital structure is one of the most difficult problems farmers face. The optimal capital structure of a farm is dependent on a number of factors including the return on assets, the cost of debt, the availability of debt and the farmer's risk preference. Sanint and Barry have modeled the optimal leverage decision using a mean-variance multiperiod risk-programming model. Even though randomness in cash flow is important, Sanint and Barry model cash flow dynamics as a linear (symmetric) response to the random variable. However, important factors in the optimal leverage decision such as liquidity risk and credit reserve risk are likely asymmetric. The asymmetry arises primarily because the consequences of a sequence of unfavorable outcomes are not simply the opposite of consequences following a sequence of favorable outcomes.

There is a need for liquidity in the normal course of a farm's operation because of the variability in farm income. Many times, however, a reduction in liquidity occurs precisely at the time when a firm needs the extra cash to meet its obligations. This is known as liquidity risk. The farmer's ability to meet cash obligations is in greatest jeopardy when a sequence of bad years occurs. If he has not carried sufficient liquidity reserves, productive assets may have to be sold off at low prices to meet cash obligations. The lower than normal prices occur because buyers of farm assets are often farmers themselves, and when it is necessary for one farmer to sell assets to cover cash obligations other farmers are likely to have similar needs. Therefore, supply may greatly exceed demand causing low prices.

Barry, Baker, and Sanint discuss forces that determine the supply of credit. These include determinants originating in the financial markets, such as monetary and fiscal policies and aggregate economic performance, along with determinants originating in agriculture, such as market conditions for commodities and land. The latter group primarily affect the lender's assessment of the farmer's credit worthiness. Credit reserve risk arises because available credit is positively correlated with expected income and asset values. Thus, holding credit reserves as a means of maintaining liquidity is risky because during those times when credit reserves are needed to be used for liquidity, the actual credit available tends to decrease.

Interest rate risk is another factor important in the leverage decision. Interest rate risk arises because farmers are not able to borrow at a constant real interest rate. Although some loans may have a fixed nominal rate, the real rate of interest varies with inflation, thus the farmer is subject to interest rate risk.

Discrete stochastic programming (DSP) is one methodology that can be used to model optimal capital structure decisions. The purpose of this paper is to illustrate the use of DSP in modeling farm capital structure. Dynamics and stochastics are modeled, not to forecast, but to capture the effects of random variable and dynamic interactions on the current decisions. Because of liquidity risk, disinvestment decisions are especially costly, therefore it is especially important to consider probability weighted future events when making investment decisions.

Modeling with Discrete Stochastic Programming

Problems modeled using DSP may be viewed as finite horizon dynamic programming problems with discrete random events and continuous choice variables. DSP is a type of mathematical programming which can model uncertainties in the right-hand sides, technical coefficients, and objective function. Because of the "curse of dimensionality", there have been only a modest number of applications of DSP since its development in 1969 by Cocks. DSP has been used to model fresh vegetable storage (Rae), investment in grain dryers and storage (Klemme), participation in government programs (Kaiser and Apland), the choice between fixed and variable loans (Leatham and Baker), the marketing of white wheat (Lambert and McCarl), and the marketing of cattle (Schroeder and Featherstone).

A schematic describing a simple DSP model is illustrated in Figure 1. The model is represented by a tree structure which is a collection of nodes and arcs with special properties. Moving forward through time corresponds to traversing the tree from left to right. Three types of nodes, random events, decisions, and payoffs represented by rectangles, bold rectangles, and at the right-hand side, terminal nodes of the tree, respectively. When a random event node is encountered, the value of the associated random variable is realized. In a DSP the random variables are assumed to have discrete distributions with each possible outcome represented by an arc to the right of the node with probability (p). When a decision node is encountered an irreversible decision is made, possibly subject to a set of constraints. This decision is made with knowledge of all past decisions and realized random variables (to the left), and knowledge of the distributions of future random variables (to the right). At the extreme right of the tree there appears a payoff (terminal) node for all realizations of the random variables. The payoff (Z) is a function of the realizations of the random variables and the choice variables encountered along that path. Thus, the terminal nodes represent the marginal probability distribution of payoffs which are functions of the decisions. The marginal probability of each payoff node is the product of the conditional probabilities of the realizations of the random variables along the associated path. For instance, in the diagram the marginal probability of the top-most payoff is $(1.0) \times (.3) \times (.4) \times (.1) = 0.012$, and the probability of the bottom-most is $(1.0) \times (.7) \times (.5) \times (.6) = .21$.

A stage is defined as the decision node and the realization of that decision. The realizations of the random variables after a decision node are called states of nature. In the schematic, subscripts refer to stages and superscripts refer to states. In contrast to dynamic programming (DP), a DSP model simultaneously computes the optimal choices for all decision nodes. Like DP all decisions are consistent with the overall problem objective. This means that each decision is affected not only by prior realizations and decisions, but also by future decisions and conditional distributions of future events.

Figure 10.1 illustrates a farm financial decision problem. The random variables are commodity prices, interest rates, and yields at different points in time, and the decision variables are how much of the commodity to raise, how much to borrow, how many assets to purchase at each of three points in time. In the terminal period, the farms net worth at the end of three periods is calculated. The sequence of events proceeds as follows. Prior to the first decision, the random variables (R_1^1) become known with certainty. The period one decisions (X_1^i) are based on knowledge of the current price $(\bar{R}_1^1)$, and the distribution of future random variables (R_2^i, R_3^i, R_4^i). Prior to the second period's decisions, the second period's random variables (R_2^1) are realized. Two values are possible for R_2^1, namely $\bar{R}_2^1$ and $\bar{\bar{R}}_2^2$ with probabilities .3 and .7, respectively. The second period decisions are made based on knowledge of the historical random variable sequence, and the future distribution of random variables. The second period's decisions may be different depending on which outcome occurred for R_2^1. The third period's decisions are chosen given the current random variables, past decisions, and the distributions of random variables in the future. Finally, the random variables in the fourth period are realized, and all the ending equity is calculated. Both revenues and profits are functions of the choice variables, X_{ij}. Thus, the tree describes a probability distribution for profits. Standard objective functions for modeling decision making under uncertainty may be used to value the payoffs.

At any stage, the distributions of future events are conditional on the realizations of the random variables up to that point. Thus, even if the realized prices $\tilde{R}_3^2$ and $\tilde{R}_3^3$ are the same, differences in the conditional distributions of future random variables may cause the decisions (X_3^2, X_3^3) to differ. Statistical dependencies such as autocorrelation in the marginal distribution of the random variables would result in these conditional distributions being different. However, if the past and future random variables are stochastically independent, then the conditional distributions are the same across all states in a given stage.

It is straightforward, in the DSP framework, to include constraints on the choice variables. For example, total debt might be limited to a fixed quantity, S. The choice of X_2^2 would then be made subject to the constraint $X_1^1 + X_2^2 \leq S$. The constraints may also include coefficients involving the random variables. For instance, the amount of capital available for purchases might be limited to C_1 in the first period, C_2 in the second period, and so on. Thus, the capital constraint for the choice of X_3^4 would be $\tilde{R}_1^1 X_1^1 + \tilde{R}_2^2 X_2^2 + \tilde{R}_3^4 X_3^4 \leq C_3$.

Since there are decision variables in every state in every stage, the number of optimal values which must be computed rises rapidly as stages and states are added. In fact, if the number of states at each stage is s and if the number of stages is k, then the corresponding schematic contains s^k decision nodes. If the number of decision variables in each state/stage combination is n and the number of constraints on those decisions is m, then the problem will contain on the order of ns^k variables and ms^k constraints. Both of these numbers grow very rapidly as the number of stages, k and the number of states, s, increase. This rapid rate of problem growth has been dubbed the curse of dimensionality.

The ability to incorporate dynamics, random events, and constraints that differ across states makes the DSP framework ideal for use in modeling liquidity risk for a farm firm. The decision variables for a DSP model focused on farm planning might include what mix of crops and livestock to produce, how much to borrow or repay, and whether to hire

labor or supply off-farm labor. Random events might include crop and livestock yields and prices (or revenues), changes in land values, and interest rates.

Notation

In order to present the mathematical model, notation is defined in this section. Variables will be in upper case and parameters will be in lower case.

Subscripts

The current state and the previous state are indicated by the following subscripts:

- t = denote the year in which a decision is made, $(t=1,...,T)$ where T is the end of the planning horizon;
- i = denote the number of the state at time $t+1$, $(i=1,...,I_{t+1})$ where I_{t+1} is the number of stages for period $t+1$; and
- j = denote the number of the state at time t, $(j=1,...,I_t)$ where I_t is the number of states for period t.

Variables

- IL = the initial amount of owned land (acres);
- IM = the initial amount of machinery owned (dollars);
- IOE = the initial amount of owner's equity (dollars);
- IHB = the initial amount of hog buildings owned (dollars);
- L_{ti} = land owned at the end of period t in state i (acres);
- PL_{ti} = land purchased at the beginning of period t in state i (acres);
- SL_{ti} = land sold at the beginning of period t and state i (acres);
- R_{ti} = Owned land planted into 1/2 corn and 1/2 soybeans at the beginning of period t in state i (acres);
- W_{ti} = owned land planted into wheat at the beginning of period t in state i (acres);
- RC_{ti} = cash rented land planted into 1/2 corn and 1/2 soybeans at period t in state i (acres);
- RS_{ti} = share rented land planted into 1/2 corn and 1/2 soybeans at period t in state i (acres);
- M_{ti} = amount of machinery owned after all period t decisions have been made in state i (dollars);
- AM_{ti} = number of acres with a machinery set at the end of period t in state i (acres);

PM_{ti} = land needing machinery purchase at the beginning of period t in state i (acres);

H_{ti} = the number of units of hogs raised during period t in state i (sows);

HB_{ti} = amount of hog buildings owned after all period t decisions have been made in state i (dollars);

PHB_{ti} = amount of hog buildings purchased at the beginning of period t in state i (dollars);

OE_{ti} = Owner's equity at the beginning of period t in state i after all decisions have been made (dollars);

D_{ti} = amount of debt at the beginning of period t in state i after all decisions have been made (dollars);

OFI_{ti} = off-farm investment during period t in state i (dollars);

OFL_{ti} = amount of good field days spent on an off-farm job during period t in state i (hours). (A good field day is a day when an operator can do field work);

SHL_{ti} = Amount of summer labor hired per good field day during period t in state i (hours); and

FHL_{ti} = amount of fall and spring labor hired per good field day during period t in state i (hours).

Technical Coefficients

Fixed technical coefficients are used to indicate the hours of good field day labor required for each of the production activities. Let:

hrs = labor needed on good field days during the summer for rotation corn and soybeans (hours);

hrf = labor needed on good field days during the fall for rotation corn and soybeans (hours);

hws = labor needed on good field days during the summer for wheat (hours);

hhs = labor needed on good field days needed for hogs during the summer (hours);

hhf = labor needed on good field days for hogs during the fall (hours);

rsf = hours used in the fall divided by hours used in the summer for off-farm employment; and

p_i = the probability of ending up at a terminal state for $i = 1,...,I_{t+1}$, where I_{t+1} is the number of terminal nodes.

Financial Coefficients

The following are the coefficients associated with the financial aspects of this model. They include coefficients used for: external credit rationing, liquidity risk, accounting, working capital and debt use, depreciation, and machinery and hog facilities requirements. Let:

dm   = one minus the rate at which machinery depreciates (%).

ma   = the value of machinery assets needed to farm the each acre of land (dollars);

fh   = the value of hog facilities needed for each sow (dollars);

db   = one minus the rate at which hog facilities depreciate (%);

pl_{ti} = price of land in period t and state i (dollars per acre);

wr   = working capital needed for rotation corn and soybeans grown on owned land (dollars per acre);

ww   = working capital needed for wheat grown on owned land (dollars per acre);

wrs   = working capital requirement for rotation corn and soybeans grown on land leased via a crop share (dollars per acre);

wrc_{ti} = working capital needed for rotation corn and soybeans grown on cash rented land for period t and state i (dollars per acre);

wh_{ti} = working capital requirement for hogs for period t in state i (dollars per sow);

tc_{ti} = transactions cost for land sale during period t in state i (dollars per acre);

pr_{ti} = after-tax profit for rotation corn and soybeans grown on owned land for period t in state i (dollars per acre);

pw_{ti} = after-tax profit for wheat grown on owned land for period t in state i (dollars per acre);

prc_{ti} = after-tax profit for cash rented rotation corn and soybean land for period t in state i (dollars per acre);

prs_{ti} = after-tax profit for share rented rotation corn and soybean land for period t in state i (dollars per acre);

ph_{ti} = after-tax profit for hogs in period t in state i (dollars per sow);

adm   = after-tax depreciation on machinery;

adf   = after-tax depreciation on hog facilities;

$rint_{ti}$ = after-tax interest rate for period t in state i paid for farm borrowing;

$rinv_{ti}$ = after-tax interest rate for period t in state i for off-farm invested capital;

aofw   = after-tax wage for off-farm employment (dollars per hour);

aplf = after-tax wage paid to fall hired labor (dollars per hour);

apls = after-tax wage paid to summer hired labor (dollars per hour);

cg_{ti} = capital gain on owned land for period t in state i (dollars per acre);

d_{ti} = the credit capacity for period t and state i in dollars (if $cg_{ti} \geq$ 0, $d_{ti} = 0$ otherwise $d_{ti} = 1$). If the capital gain on land is negative, the current debt limit is equal to last period's debt limit;

e_{ti} = the weighing for credit capacity for period t and state i (if cg_{ti} < 0 then $e_{ti} = 0$ otherwise e_{ti} is positive); and

Θ = the "risk aversion" coefficient. $\Theta \geq 0$ indicating that the individual is risk neutral ($\Theta = 0$) or risk averse ($\Theta > 0$). Θ is the Pratt-Arrow measure of relative risk aversion.

Right-hand Side Coefficients

bl = the initial amount of land owned by the decision maker (acres);

bm = the initial amount of machinery owned by the decision maker (dollars);

bhb = the initial amount of hog buildings owned by the decision maker (dollars);

be = the decision maker's beginning owner's equity (dollars);

s = the maximum amount of good field time available during the summer (hours);

f = the maximum amount of good field time available during the fall (hours);

c = the amount of withdrawals made by the family for living expenses (dollars);

bofl = the maximum amount of off-farm labor allowed (hours);

shlb = the maximum amount of summer hired labor (hours).

fhlb = the maximum amount of fall hired labor (hours); and

wb = the maximum amount of wheat grown on owned land. This is the amount of wheat base that a farmer has under government programs.

The Mathematical Model

A mathematical model of a representative Indiana corn-soybean-hog farm consistent with the above notation is described in this section.

Objective Function

(1)   maximize $u = \sum_{i=1}^{I_{t,1}} p_i \dfrac{(OE_{T,1})^{1-\theta}}{1-\theta}$

The objective is to maximize the expected utility (u) of terminal net worth where the farmers utility function is assumed to be a power utility function. Featherstone, Preckel, and Baker discuss the power utility function and suggest a method to evaluate negative terminal net worth using a power function. The expected utility of the decision maker is maximized subject to the following sets of constraints.

Owned Land Usage Constraints

(2a) $-IL - PL_{11} + SL_{11} + R_{11} + W_{11} \leq 0$
(2b) $-L_{t\text{-}1j} - PL_{ti} + SL_{ti} + R_{ti} + W_{ti} \leq 0$

for $t = 1,...,T$ and $j = 1,...,I_{t\text{-}1}$. Given j, i goes from $(j\text{-}1) * I_t/I_{t\text{-}1} + 1$ to $j*I_t/I_{t\text{-}1}$.[1] These constraints are used to ensure that owned land sold or used for rotation corn and soybeans or wheat is less than or equal to land previously owned on newly purchased.

Owned Land Accounting Constraints

(3a) $- IL - PL_{11} + SL_{11} + L_{11} = 0$
(3b) $- L_{t\text{-}1j} - PL_{ti} + SL_{ti} + L_{ti} = 0$

for $t = 2,...,T$ and $j = 1,...,I_{t\text{-}1}$. Given j, i goes from $(j\text{-}1) * I_t/I_{t\text{-}1} + 1$ to $j*I_t/I_{t\text{-}1}$.
The owned land accounting constraints are used to transfer owned land to the next period. These constraints require that the land owned last period plus purchases minus land sales is equal to ending owned land.

Acres with Machinery Accounting Constraints

(4a) $- IL - PM_{11} + AM_{11} = 0$
(4b) $- dm\ AM_{t\text{-}1j} - PM_{ti} + AM_{ti} = 0$

for $t = 2,...,T$ and $j = 1,...,I_{t\text{-}1}$. Given j, i goes from $(j\text{-}1) * I_t/I_{t\text{-}1} + 1$ to $j*I_t/I_{t\text{-}1}$.
These constraints keep track of the machinery capacity. These constraints are necessary because machinery depreciates. Last year's

machinery is only able to service a portion (one minus the machinery depreciation rate) of the land it was able to last year.

Machinery Assets Accounting Constraints

(5a) $\quad - IM - ma\ PM_{11} + M_{11} = 0$

(5b) $\quad - dm\ M_{t-1j} - ma\ PM_{ti} + M_{ti} = 0$

for $t = 2,...,T$ and $j = 1,...,I_{t-1}$. Given j, i goes from $(j-1) * I_t/I_{t-1} + 1$ to $j*I_t/I_{t-1}$.

These constraints transfer the value of machinery from year to year. The depreciated value of last year's machinery plus the value of new machinery purchases is equal to machinery transferred to the next period.

Machinery Purchases Constraints

(6a) $\quad - IL + R_{11} + W_{11} + RC_{11} + RS_{11} - PM_{11} \leq 0$

(6b) $\quad - dm\ AM_{t-1j} + R_{ti} + W_{ti} + RC_{ti} + RS_{ti} - PM_{ti} \leq 0$

for $t = 2,...,T$ and $j = 1,...,I_{t-1}$. Given j, i goes from $(j-1) * I_t/I_{t-1} + 1$ to $j*I_t/I_{t-1}$.

These constraints set the machinery purchases. The depreciated machinery set from the previous year plus new purchases must be sufficient to service the acreage farmed.

Hog Facilities Purchases Constraints

(7a) $\quad - IHB + fh\ H_{11} - PHB_{11} \leq 0$

(7b) $\quad - db\ HB_{t-1j} + fh\ H_{ti} - PHB_{ti} \leq 0$

for $t = 2,...,T$ and $j = 1,...,I_{t-1}$. Given j, i goes from $(j-1) * I_t/I_{t-1} + 1$ to $j*I_t/I_{t-1}$.

These constraints set the amount of hog facilities to be purchased. The depreciated amount of facilities plus new hog facility purchases must be greater than the number of sows the farmers intends to raise this year.

Hog Assets Accounting Constraints

(8a) $\quad - IHB - PHB_{11} + HB_{11} = 0$

(8b) $\quad - db\ HB_{t-1j} - PHB_{ti} + HB_{ti} = 0$

for $t = 2,...,T$ and $j = 1,...,I_{t-1}$. Given j, i goes from $(j-1) * I_t/I_{t-1} + 1$ to $j*I_t/I_{t-1}$.

These constraints transfer the value of hog facilities from year to year. The depreciated value of last year's hog facilities plus the value of new hog facility purchases is equal to hog facilities transferred to the next period.

Debt Limit Constraints

(9a) $- e_{t,i} OE_{11} + D_{11} \leq 0$
(9b) $- d_{ti} D_{t\text{-}1j} - e_{ti} OE_{ti} + D_{ti} \leq 0$

for $t = 1,...,T$ and $j = 1,...,I_{t\text{-}1}$. Given j, i goes from $(j\text{-}1) * I_t/I_{t\text{-}1} + 1$ to $j*I_t/I_{t\text{-}1}$.

These constraints represent the external credit rationing from the lender. If last year's capital gain associated with land was positive, then the amount of debt available to be borrowed is equal to a proportion of owner's equity. If, however, the capital gain associated with land is negative, then the maximum amount of credit available to this farm operator is the amount of debt held last year.

Summer Labor Constraints

(10)   hrs R_{ti} + hws W_{ti} + hrs RC_{ti} + hrs RS_{ti} + OFL_{ti} + hhs H_{ti}
        - $SHL_{ti} \leq s$

for $t = 1,...,T$ and $i = 1,...,I_t$.

These constraints ensure that labor used for 10 weeks during the summer is not more than or equal to that available. The amount of labor used on good field days during the summer for crops and the hog enterprises plus the amount hired out must be less than labor available from the permanent labor force plus the summer labor hired in.

Spring and Fall Labor Constraints

(11)   hrf R_{ti} + hrf RC_{ti} + hrf RS_{ti} + rsf OFL_{ti} + hhf H_{ti}
        - $FHL_{ti} \leq f$

for $t = 1,...,T$ and $i = 1,...,I_t$.

These constraints require that spring and fall labor hired for 24 weeks on good field days will not exceed what is available. The amount of spring and fall labor used for crops, hogs, and off-farm employment must be less than or equal to labor available from the permanent labor force plus the fall labor hired in.

Balance Sheet Constraints

(12) $pl_{ti} L_{ti} + HB_{ti} + M_{ti} + wr R_{ti} + ww W_{ti} + wrc_{ti} RC_{ti} + wrs RS_{ti}$
$+ wh_{ti} H_{ti} + OFI_{ti} - OE_{ti} - D_{ti} = 0$

for $t = 1,...,T$ and $i = 1,...,I_t$.

These constraints are accounting constraints which simply say that assets equal debt plus owners equity. These constraints also set the amount of debt.

Owner's Equity Constraints

(13a) $IOE - tc_{11} SL_{11} - OE_{11} = 0$

(13b) $pr_{ti} R_{t-1j} + pw_{ti} W_{t-1j} + prc_{ti} RC_{t-1j} + prs_{ti} RS_{t-1j} + ph_{ti} H_{t-1j}$
$- adm M_{t-1j} - adf HB_{t-1j} - rint_{ti} D_{t-1j} + rinv_{ti} OFI_{t-1j}$
$+ aofw OFL_{t-1j} - aplf FHL_{t-1j} - apls SHL_{t-1j} + OE_{t-1j}$
$+ cg_{ti} L_{t-1j} - tc_{ti} SL_{ti} - OE_{ti} = c$

for $t = 2,...,T+1$ and $j = 1,...,I_t$. Given j, i goes from $(j-1) * I_t/I_{t-1} + 1$ to j^*I_t/I_{t-1}.

These constraints calculate the owner's equity. Owner's equity is the income from both farm and nonfarm activities such as crops, hogs, off-farm employment, and off-farm investment minus expenses such as hired labor expense, depreciation, and interest expense minus a withdrawal for family living and consumption plus last years ending owner's equity. Included in this calculation of owner's equity is an adjustment for a change in the market value of land (cg_{ti}). The adjustment in land value is included to calculate market value net worth.

Initial Period Bounds

(14a) $IL = bl$
(14b) $IM = bm$
(14c) $IHB = bhb$
(14d) $IOE = be$

These constraints set the initial conditions for the farm operation. Initial amounts of land, machinery, owner's equity and hog buildings are all specified.

Other Bounds

(15a) $OFL_{ti} \leq bofl$ for $t = 1,...,T$ and $i = 1,...,I_t$.
(15b) $SHL_{ti} \leq shlb$ for $t = 1,...,T$ and $i = 1,...,I_t$.

(15c) $FHL_{ti} \leq fhlb$ for $t = 1,...,T$ and $i = 1,...,I_t$.

(15d) $W_{ti} \leq wb$ for $t = 1,...,T$ and $i = 1,...,I_t$.

These constraints place an upper limit on the off-farm employment, the hired labor and the wheat activities. The labor bounds are needed to limit the size of the farming operation. The wheat bound is needed when farm programs are in effect as this bound represents the amount of wheat base the farm operator has on which he is eligible for deficiency payments.

Stochastic Environment

The above mathematical model allows for many stochastic variables. These include prices, yields, land values, cash rents and interest rates. This section describes the procedures used to define the stochastic environment. The real interest rate is assumed to follow a stochastic process that is independent of all other prices. That is:

(16) $r_t = r_{t-1} + e_t,$

where r_t is the real interest rate for year t, e_t is the random component; it is assumed to be distributed normally with a zero mean and standard deviation equal to .02 based on 1960 through 1985 data.

Gallagher provides evidence that U.S. corn yields are negatively skewed with an upper limit on output and occasional low yields. Based on Gallagher's work, it is assumed that corn yield has a negatively skewed distribution. It is also assumed that soybean and wheat yields are negatively skewed. The stochastic nature of corn yield is modeled in the following manner:

(17) $Y_{Ct} = M_C + b_1 t - e_{Ct},$

where Y_{Ct} is the yield for corn in year t, M_C is the maximum potential yield for corn, which is assumed to be 20% greater than the maximum detrended yield observed from 1960 through 1984 in district #4 in Indiana, b_1 is the estimated time trend in yield, and e_{Ct} is the detrended yield deviation from maximum. The yields of soybeans and wheat are modeled in the same manner. The error terms on the corn, soybean, and wheat yield equations are assumed to be distributed multivariate lognormal with mean μ and variance Σ. Data from 1960 to 1984 are used to estimate μ and Σ.

Corn, soybeans, wheat, and hog direct production costs as well as tax rates, wages, consumption, machinery prices and hog facility prices are assumed to be nonstochastic for this study. Machinery complements chosen and further information can be found in Featherstone.

Policy Environment

This section describes the policy environment and the modeling of variables affected by government farm programs. This study examines optimal capital structure under loan rates and target prices at the 1985-86 level. PC-WHEATSIM and FEEDSIM models (Holland and Sharples; Chattin, Hillberg, and Holland) were used to stochastically simulate corn, soybean, and wheat prices under the 1985 programs. These models were chosen because they are designed to perform stochastic simulations of alternative farm programs.

To facilitate input into the mathematical programming model, the results from PC-WHEATSIM and FEEDSIM were further transformed. For this study, it is assumed that the decision maker perceives corn, soybean and wheat prices to follow a stochastic process in which the price of corn is a sum of the previous years price, plus a mean change in price, and a normally distributed zero-mean random error term. The data used to calculate the mean change and the random error term were the data generated by FEEDSIM and PCWHEATSIM.

The price of hogs is assumed to follow a similar stochastic process. Annual hog price is affected by the previous year's hog price, the previous years corn price, an expected change in hog price, and a zero mean random error process. More formally, hog prices are assumed to have the following stochastic process:

$$(18) \qquad P_{Ht} = 19.61 + .39\ P_{Ht-1} + 4.52\ \bar{P}_{Ct-1} + \Delta\bar{P}_{Ht} + e_{Ht}$$
$$\phantom{(18) \qquad P_{Ht} = } (1.69) \quad\quad (1.79) \quad\quad\ (1.24)$$

where P_{Ht} represents the price of hogs in period t, $\Delta\bar{P}_{Ht}$ represents the mean change in price in year t, and e_{Ht} represents the random shock to hog prices in year t. Equation 18 was estimated using Indiana data from 1960 through 1983. The values in parentheses below the parameter estimates are T-values. The random errors of the corn price, soybean price, the wheat price, and the hog price equations are distributed multivariate normal with a mean of zero.

Agricultural policies are assumed to affect cash rent via the effect on crop prices and affect land values through cash rent (Featherstone and Baker). Equations for cash rent and land price were estimated using historical nominal returns, cash rent, and land values in Indiana from 1960 to 1984. Cash rent is a function of past cash rent and the residual returns to an acre of corn and soybeans grown in rotation. In the short-run, a one dollar change in the real returns to rotation corn and soybeans will raise real cash rent by 8.1 cents. In the long-run, a one dollar change in returns will increase cash rent by 60.3 cents.

The modeling of land price under alternative policies is based upon the familiar capitalization formula. Land price is a function of rent and lagged land prices. Land price exhibits cyclical behavior similar to that documented by Burt. In the short-run, a one dollar increase in cash rent is capitalized at the real rate of 16.8%. In the long run, a one dollar increase in rent is capitalized at the real rate of 5.7%.

The probability distributions specified above are continuous. However, discrete outcomes are necessary for the DSP model. A multivariate discrete approximation was made for the multivariate continuous probability space. The method used for this study divided the multivariate continuous probability space into regions. For each region, the probability of being in the region and the conditional means of the region were calculated using the numerical integration technique discussed in Kaylen and Preckel. The vector of conditional means in each region was then chosen as the discrete outcomes.

The farm model was chosen to be a four year model with 900 terminal nodes. The model has 4,262 constraints, 6,225 activities, and 900 nonlinear variables. Nine states of nature were chosen for the first year, 5 states for the second and third years and 4 states for the fourth year. The model was solved using MINOS 5.0 on a IBM 3083 mainframe computer. Model solution required 8 megabytes of memory. Moving from the expected wealth maximizing solution (LP) to the Pratt-Arrow relative risk aversion level of 3.75 required 11 minutes and 11 seconds of CPU time. The time required to obtain the linear programming solution was roughly 2 hours of CPU time. The cost of each CPU hour was $500.

Results

The farm level effects of the 1985 programs are examined for a representative midwestern corn-soybean-hog farm under two initial leverage positions (40% and 70% debt to assets). The initial conditions were assumed to be those farmers were facing at the beginning of 1986. The models were identical except for the initial amounts of debt, equity and credit available. The 40% debt farm has initial equity of $750,000, initial debt of $500,000 and a 50% debt-to-asset credit limit. The 70% farm has $375,000 of initial equity, $875,000 of initial debt, and a 75% debt-to-asset credit limit. For each initial debt position, the effect of risk aversion on the optimal portfolio of investments is examined. Eight levels of risk aversion are studied ranging from a Pratt-Arrow measure of relative risk aversion of 22.5 to .0375. The range is based on McCarl's suggestion that the absolute risk aversion coefficient should fall between 0 and 10 divided by the standard error of income.

The expected terminal net worth, the standard deviation in terminal wealth, the coefficient of variation of terminal wealth, and the certainty equivalent of terminal wealth for a 40% and 70% debt farm is found in Table 10.1. Under both farms, terminal wealth and the certainty equivalent of ending terminal wealth is expected to increase. The coefficient of variation of terminal wealth of the 70% debt farm is about twice that of the 40% debt farm. The farmer with 40% debt is earning a certainty equivalent real annual rate of return of between 7.2% and 13.4% depending on his level of risk aversion.[2]

Table 10.1 Mean, Standard Deviation, Coefficient of Variation, and Certainty Equivalent of Ending Wealth on a Farm with 40% and 70% Initial Debt Under the 1985 Commodity Programs

Θ[a]	Mean	Standard Deviation	Coefficient of Variation	Ceratinty Equivalent
	----thousands of dollars---		percent	thousand dollars
		40% Debt		
.0375	1242	325	26.2	1241
.075	1242	325	26.2	1239
.375	1242	322	25.9	1228
.75	1242	322	25.9	1215
3.75	1241	319	25.7	1137
7.5	1238	315	25.4	1080
17.5	1229	310	25.3	1008
22.5	1227	310	25.3	989
		70% Debt		
.0375	837	415	49.6	834
.075	837	415	49.6	830
.375	836	411	49.2	806
.75	836	410	49.0	778
3.75	832	406	48.8	633
7.5	819	394	48.1	554
17.5	783	382	48.7	474
22.5	781	385	49.3	453

[a] Θ is the Pratt-Arrow relative risk aversion coefficient.

The farmer operating with initial debt of 70% earns a certainty equivalent rate of return of between 4.8% and 22.1%.

If the farmer is nearly risk neutral (.0375 or .075) and operating with 40% debt, he increases leverage to the limit of 50% (Table 10.2). However, the extra money borrowed during the first year is invested off the farm and used for investment after the first year. If the most likely state of nature occurs, land price decreases and the farmer is unable to borrow more than he had the previous year. Because of this, the farmer borrows the first year when the credit is available and invests the money to maintain the liquidity reserve for the following year at a cost of 2%. If the farmer is more risk averse, the farmer does not deviate from the initial debt to asset ratio other than to borrow an additional amount for operating capital. The farmer with 70% debt and a risk aversion coefficient of .0375, .075, or .375 increases borrowing and holds the borrowed money in off-farm investment. However, in no case does the farmer with 70% debt deplete all credit reserves. If the farmer is risk averse (17.5 or 22.5), then borrowing is reduced by selling off land.

The first year's portfolio of hogs, rotation acreage, and wheat are identical for the 40% and the 70% at every level of risk aversion except 17.5 and 22.5 (Table 10.3). The farm with 40% debt is larger the second year than the farm with 70% debt. As the farmer becomes more risk averse, business risk and financial risk are reduced in the second year. Hogs are only raised in the initial facilities; no new investment into hog facilities occurs. Shadow prices and penalty costs expressed in certainty equivalents are presented in Table 10.4.[3]

Another acre of land available for use at no cost to the operator has a greater certain return for the farmer with 70% debt than with 40% debt for all but one risk aversion level (7.5). This may occur because the farmer needs the extra cash flow to meet payments. Thus, being able to obtain an extra means to generate cash flow for the farmer 70% initial debt at no cost is more valuable. Cash renting has a lower penalty cost on the 40% farm than on the 70% farm for all levels of risk aversion. This occurs because cash rent can be viewed as another means of debt, i.e., a fixed payment is due regardless of the revenue generated. Share renting has a larger penalty cost for the more risk neutral farmer (.0375 to .75) under 70% debt. A very risk averse farmer (17.5, 22.5) with 70% debt raises 64 acres of share rented land.

It is assumed that the farmer has a 100 acre wheat base. The wheat base is constraining under both 40% and 70% initial debt for the first year (Table 10.4). As a farmer becomes more risk averse, another acre of wheat base becomes more valuable under both debt scenarios. Diversification activities are more valuable on the farm with 70% debt than the farm with 40% debt. At every level of risk aversion, the 70% farmer is willing to pay more for diversification activities.

Table 10.2   Debt, Credit Reserves, and Off-Farm Investment on a Farm with 40% and 70% Initial Debt Under the 1985 Commodity Programs

Θ[a]	Debt Assets	----Year 1----			----Year 2[b]		
		Debt	Credit Reserves	Off-farm Investment	Debt	Credit Reserves	Off-farm Investment
	%	---------thousands of dollars---------					
		40% Debt					
.0375	50.0	750	0	240	750	0	0
.075	50.0	750	0	240	750	0	0
.375	40.5	510	240	0	510	0	0
.75	40.5	510	240	0	510	0	0
3.75	40.5	510	240	0	510	0	0
7.5	40.5	510	240	0	478	32	0
17.5	40.5	510	240	0	478	32	0
22.5	40.5	510	240	0	445	65	0
		70% Debt					
.0375	74.7	1109	16	224	1109	0	0
.075	74.7	1109	16	224	1109	0	0
.375	70.5	895	230	10	895	0	0
.75	70.2	885	240	0	885	0	0
3.75	70.2	885	240	0	872	13	0
7.5	70.2	885	240	0	863	22	0
17.5	68.2	782	310	0	728	55	0
22.5	68.2	782	310	0	728	53	0

[a] Θ is the Pratt-Arrow relative risk aversion coefficient.
[b] The second year decisions listed here have a 1 in 3 chance of occuring and is from the most second year state of nature.

Table 10.3 Crop and Hog Diversification on a Farm with 40% and 70% Initial Debt Under the 1985 Commodity Programs

| θ[a] | ---Year 1--- | | | ---Year 2[b] | | |
	Debt	Credit Reserves	Off-farm Investment	Debt	Credit Reserves	Off-farm Investment
			40% Debt			
.0375	700	100	51	977	0	45
.075	700	100	51	977	0	45
.375	700	100	51	818	0	45
.75	700	100	51	818	0	45
3.75	700	100	51	721	100	45
7.5	700	100	51	700	100	45
17.5	700	100	51	700	100	45
22.5	700	100	51	623	100	45
			70% Debt			
.0375	700	100	51	954	0	45
.075	700	100	51	954	0	45
.375	700	100	51	813	0	45
.75	700	100	51	806	0	45
3.75	700	100	51	700	100	45
7.5	700	100	51	679	100	45
17.5	596	100	51	596	100	45
22.5	596	100	51	596	100	45

The incentives for growth in farm size during year 2 are greater if debt is initially 40% rather the 70% (Table 10.5). However, under the most likely state of nature, after three years, the risk neutral farmer (.0375 or .075) will own more land under the 70% debt scenario than under 40% debt. The more risk averse farmer (17.5 or 22.5) will reduce land holdings under the 70% debt scenario, while maintaining land holdings under the 40% debt scenario.

A farmer operating under the 70% debt scenario is willing to pay more for another acre of land at risk aversion levels of .75 or less (Table 10.5, column 5). If the farmer is more risk averse (≥ 3.75), another acre of land is worth more under the 40% debt scenario. Under both scenarios, if a farmer is more risk averse, he is willing to offer less to get another acre of land. The marginal value of another acre of land is negative for the more risk averse farmer (17.5 or 22.5) under the 70% debt scenario.

A farmer operating with 70% debt is willing to pay the most for an additional dollar of beginning equity (Table 10.5, columns 6 and 7). A farmer that is nearly risk neutral (.0375) earns a certainty equivalent annual return to initial equity of 10.4% and 14.9% at the margin under the 40% and 70% debt scenarios, respectively. A farmer that is more risk averse (3.75) earns 9.1% and 11.8% at the margin under the 40% and 70% debt scenarios, respectively. At all levels of risk aversion, the marginal rate of return on an extra dollar of equity is greater for the farmer with higher debt levels as would be expected.

Table 10.4 Marginal Contribution of Crops and Hogs to Certainty Equivalents for a Farm with 40% and 70% Initial Debt Under the 1985 Commodity Programs

Θ^a	Economic Rent on Owned Land[b]	Cash Rented Land	Share Rented Land	Wheat Base	Hog Facilities
		40% Debt			
.0375	73.26[c]	-22.39[c]	-16.53[c]	27.92[c]	.196[d]
.075	73.37	-22.20	-16.27	28.06	.196
.375	75.64	-19.49	-12.98	29.26	.197
.75	78.70	-15.86	-8.62	30.66	.197
3.75	79.57	-11.63	0	38.84	.195
7.5	74.51	-14.23	0	43.80	.196
17.5	70.24	-16.31	0	47.95	.199
22.5	69.61	-16.64	0	48.59	.197
		70% Debt			
.0375	79.64	-32.86	-27.48	29.82	.224
.075	79.88	-32.30	-26.78	30.09	.224
.375	83.47	-26.40	-19.63	32.48	.224
.75	89.97	-17.82	-9.44	35.29	.223
3.75	84.11	-16.05	0.00	48.73	.219
7.5	48.10	-50.89	-31.61	55.09	.241
17.5	77.42	-22.82	0[e]	62.06	.294
22.5	76.62	-24.20	0[e]	65.16	.325

[a] Θ is the Pratt-Arrow relative risk aversion coefficient.
[b] The economic rent is from the cropland accounting constraint in the first year. The cash rented land is the penalty cost associated for cash rented land. The share rented land is the penalty cost associated with a 50-50 share rent lease.
[c] dollars of certainty equivalents per acre.
[d] dollars of certainty equivalents per dollar of hog facilities.
[e] 64 acres of corn-soybean rotation are share rented.

Table 10.5 Acres and Marginal Value of Owned Land and Marginal Value of an Initial Equity on the Indiana Corn-Soybean Farm with 40% and 70% Initial Debt Under the 1985 Commodity Programs

Θ^a	Owned Land [b]			Shadow Value of Land	Shadow Value of Initial Equity (growth)[d]
	Year 1	Year 2[c]	Year 3[c]		
			40% Debt		
.0375	800	977	1063	206.63	1.484 (10.4)
.075	800	977	1063	206.06	1.483 (10.3)
.375	800	818	1064	200.49	1.476 (10.2)
.75	800	818	1064	193.48	1.468 (10.1)
3.75	800	821	821	146.62	1.419 (9.1)
7.5	800	800	800	109.62	1.383 (8.4)
17.5	800	800	800	53.38	1.349 (7.8)
22.5	800	800	800	39.14	1.346 (7.7)
			70% Debt		
.0375	800	954	1069	231.62	1.745 (14.9)
.075	800	954	1069	230.33	1.740 (14.9)
.375	800	813	1020	218.61	1.705 (14.3)
.75	800	806	868	204.40	1.674 (13.7)
3.75	800	800	808	113.27	1.563 (11.8)
7.5	800	800	800	39.98	1.547 (11.5)
17.5	696	696	606	-46.09	1.569 (11.9)
22.5	696	696	606	-66.03	1.579 (12.1)

[a] Θ is the Pratt-Arrow relative risk aversion coefficient.

[b] The initial amount of owned land is 800 acres.

[c] The second year decisions listed here have a 1 in 3 chance of occurring and is from the most second year state of nature. The third year decisions listed here have a 1 in 6 chance of occurring and is from the most likely third year state of nature.

[d] The growth rate (%) is implied by the shadow value of equity. Calculated by taking the fourth root of the shadow value.

Conclusions

The modeling of optimal capital structure has confronted agricultural economists for many years. This paper has demonstrated the use of discrete stochastic programming (DSP) to model optimal capital structure. DSP was chosen because of its ability to handle factors such as liquidity risk and credit reserve risk. These are difficult to handle because the consequence of a sequence of unfavorable outcomes are not simply the opposite of consequences following a sequences of favorable outcomes. Direct expected utility maximization using a nonlinear utility function can also be handled in the DSP framework. DSP was also used because of the ease that shadow prices can be calculated and interpreted. This allows the modeler to draw important inferences which may be unavailable by simply examining the primal solution.

In the process of building the DSP model, several issues were raised that deserve further research. Perhaps the most critical is the discrete approximation to the probability distributions. The probability distributions were constructed in a manner such that the means were equal. A recent paper by Miller and Rice demonstrates that in the univariate case using the conditional mean and the associated probability will underestimate the variance of the distribution. Other moments of the distributions such as skewness and kurtosis are also underestimated. They suggest an alternative procedure based on Gaussian quadrature. In the case of subjective, continuous distributions, Gaussian quadrature must be used twice, once to calculate the subjective distributions, and once to calculate the discrete distribution. Generalizing the technique to the multivariate case would help the modeler to have more confidence that the selection of the discrete states would not bias the results in an unknown direction.

Another area of study needing further study is the choice of an appropriate objective function. In this paper, it was assumed that the maximization of expected utility of terminal net worth was appropriate subject to the subtraction of a fixed consumption amount. Do farmers really behave this way or do they adjust consumption based on changes in net worth? A final area of study would be the appropriate length of planning horizon to get "realistic" first period decisions when considering optimal capital structure choice. This paper used a four year model. Although the "curse of dimensionality" is retreating this will still be a relevant question for DSP users. There will always be a tradeoff between the quality of the probability information and the length of the planning horizon. However, DSP is a viable alternative for modeling dynamic and stochastic processes.

Notes

1.   It is important to get the indices correct when connecting different states of nature across time in a DSP. It is straightforward to compute the indices of the arcs emanating from a given node because at a given stage, the number of arcs per node is equal across states. The computation of the indices is based on three things: the index of the arc traversed in the previous period (j), the total number of nodes in the current period (I_t), and the total number of nodes in the previous period (I_{t-1}).

2.   The risk free rate of return is calculated by $\left(\dfrac{CE}{\text{Initial Wealth}} \right)^{1/4} - 1$, where CE is the certainty equivalent.

3.   As with many dual variables, these shadow prices may or may not be unique (Paris).

References

Barry, P.J., C.B. Baker, and L.R. Sanint. "Farmers' Credit Risks and Liquidity Management." *American Journal of Agriculture Economics* 63(1981):216-27.

Burt, O.R. "Econometric Modeling of the Capitalization Formula for Farmland Prices." *American Journal of Agriculture Economics* 68(1986):10-26.

Chattin, B.L., A.M. Hillberg, and F.D. Holland. *PC WHEATSIM: Model Description and Computer Program Documentation*. Station Bulletin No. 477, Dept. of Agr. Econ., Agr. Exp. Sta., Purdue University, September 1985.

Cocks, K.D. "Discrete Stochastic Programming." *Management Science* 15(1968):72-79.

Featherstone, A.M. "A Portfolio Choice Model of the Financial Response of Indiana Farms to Alternative Price and Income Support Programs." Unpublished Ph.D. Thesis, Purdue University, August 1986.

Featherstone, A.M., P.V. Preckel, and T.G. Baker, "Incorporating Liquidity into the Financial Modeling of the Farm Firm." Presented at the Society of Economic Dynamics and Control Annual Meetings, Tempe, Arizona, March 9, 1988.

Featherstone, A.M. and T.G. Baker, "The Effect of Reduced Price, and Income Supports on Farmland Rent and Value." *North Central Journal of Agricultural Economics* 10(1988):177-89.

Gallagher, Paul. "U.S. Corn Yield Capacity and Probability: Estimation and Forecasting with Nonsymmetric Disturbances." *North Central Journal of Agricultural Economics* 8(1986):109-22.

Holland, F.D., and J.A. Sharples. *FEEDSIM: Model 15 Description and Computer Program Documentation*. Station Bulletin No. 387, Dept. of Agr. Econ., Agr. Exp. Sta., Purdue University, August 1982.

Kaiser, H.M. and Jeffery Apland. *A Risk Analysis of Farm Program Participation*. Station Bulletin 578. Department of Agricultural and Applied Economics, University of Minnesota, 1987.

Kaylen, M.S. and P.V. Preckel. "MINTDF: A FORTRAN Subroutine for Computing Parametric Integration." Station Bulletin #519, Agricultural Experiment Station, Purdue University, May 1987.

Klemme, R.M. "An Economic Analysis of the On-Farm Grain Handling Decision Problem." Unpublished Ph.D. Thesis, Purdue University, May 1980.

Lambert, D.K. and B.A. McCarl, "Sequential Modeling of White Wheat Marketing Strategies." *North Central Journal of Agricultural Economics* 11(1989):105-15.

Leatham, D.J. and T.G. Baker. "Farmers' Choice Fixed and Adjustable Rate Loans." *American Journal of Agriculture Economics* 70(1988):803-12.

McCarl, B.A. "Innovations in Programming Techniques for Risk Analysis." *Risk Analysis for Agricultural Production Firms: Concepts, Information Requirements and Policy Issues.* Proceedings of Southern Regional Project S-180. Department of Agricultural Economics, Agricultural Experiment Station, Washington State University at Pullman, Washington, August 1986, pages 94-111.

Miller III, A.C. and T.R. Rice. "Discrete Approximations of Probability Distributions," *Management Science* 29(1983):352-62.

Paris, Quirino. "Multiple Optimal Solutions in Linear Programming Models." *American Journal of Agriculture Economics* 63(1981):724-7.

Rae, A.N., "An Empirical Application and Evaluation of Discrete Stochastic Programming in Farm Management." *American Journal of Agriculture Economics* 53(1971):625-38.

Sanint, L.R., and P.J. Barry. "A Programming Analysis of Farmers' Credit Risks." *American Journal of Agriculture Economics* 65(1983):321-5.

Schroeder, T.C. and A.M. Featherstone. "Dynamic Marketing and Retention Decisions for Cow-Calf Producers." *American Journal of Agriculture Economics* 72,4(1990):1028-40.

11

A Dynamic Programming Analysis of a Variable Amortization Loan Plan

Gary D. Schnitkey and Frank S. Novak

Returns to agricultural assets are risky and are generally viewed as being dynamic. While returns vary, obligations on debt instruments traditionally have remained constant over time. Constant debt obligations tend to destabilize the farm's financial position, because loan obligations are not positively correlated with returns to assets. If loan terms allow for a positive correlation between returns to assets and debt obligations, the firm's financial position may be enhanced. Various "innovative" loan instruments which allow for this positive correlation have been proposed (see Lee and Baker for a review).

The purpose of this paper is to analyze the performance of a variable amortization loan repayment plan. This plan has been proposed by Baker (1976, 1986) and is unique in that it allows for a debt reserve, which provides liquidity in periods of adverse incomes. The variable amortization loan plan's performance is compared to a straight amortization loan and a flexible amortization loan. These alternative loan plans are described in the next section.

In comparing the three plans, dynamic programming models are used. These programs model an investment in a hog finishing operation under differing loan repayment plans and loan levels, considering the conflicting objectives of borrowers and lenders. The models determine the optimal decisions rules for withdrawing funds from the operation and the yearly repayment on principal. Criteria for evaluating alternative loan plans include the future value of withdrawals from the hog finishing operation, the probability of suffering a cash shortfall, and the amount of debt outstanding at the end of the loan term.

Alternative Loan Repayment Plans

One of the more common loan repayment plans is a traditional amortization loan. Under this plan, total periodic loan obligations remain constant over the loan's life. Thus, over time, the principal and interest portions of the total payment vary, with the interest payment constituting a larger portion of the total payment in earlier years than in later years. The advantage of this plan is simplicity. A disadvantage is that loan repayments do not vary with returns to assets.

Another loan repayment plan is what is termed here as a flexible amortization plan. This plan is offered by the Farm Credit System, and is essentially equivalent to a plan proposed by Lee. Under this plan, total debt obligations are calculated as under the amortization loan. In each year, the principal must be reduced to the level prescribed by the amortization schedule. In any given year, farmers can further reduce the principal portion of the loan. Reducing principal then reduces future interest payments. In a future year, a principal payment may not have to be made if the principal has been previously paid.

One advantage of the flexible amortization plan over the traditional amortization plan is that farmers can repay future principal payments in years of high income. Thus, the plan provides some flexibility in matching returns from assets to loan repayments. If, however, principal has not been prepaid, the plan does not allow for lower debt repayments in adverse income years.

An alternative to the flexible amortization plan is Baker's variable amortization plan. This plan includes two key elements. First, a debt reserve is added to the total loan obligation. This debt reserve, equaling a certain percentage of the total loan, can be drawn on in adverse income years. For example, a 10 percent debt reserve could be added to the loan. This means that a $4,000 reserve exists on a $40,000 loan. The second key component involves scheduling of loan repayments. According to Baker, loan debt obligations can be determined using an amortization schedule based on the loan plus the debt reserve. Then, the loan obligation would be allowed to "flex" based on returns to assets. The repayment during a year could be determined based on an index of revenues and costs (see Baker (1986) for an example and further description). An advantage of the variable amortization plan is that it allows loan repayments to be positively correlated to return on assets. It also allows more flexibility than does the flexible amortization plan.

Evaluating Alternative Loan Repayment Plans

When evaluating the alternative loan repayment plans, two perspectives are considered: the borrower's (i.e., farmer's) perspective and lender's perspective. Profit-maximizing, risk neutral borrowers prefer loan plans with lower interest costs and allow for faster withdrawals from the debt-financed investment. In the case of a farm, withdrawals of funds allow for firm growth, investment in off-farm assets, and family living expenses. An objective which incorporates this perspective is the maximization of the present value of withdrawals.

Lenders, on the other hand, prefer plans that have the highest probability of meeting loan obligations. In addition, lenders prefer that borrowers maintain liquid funds, such as in a savings account. This liquidity source then can be used to meet loan obligations in adverse income years. An objective which minimizes the discounted present value of cash shortfalls can capture this perspective.[1]

The Stochastic Dynamic Programming Models

Stochastic dynamic programming (DP) models are solved to evaluate the three alternative loan repayment plans. Results from the models are generated for three differing perspectives: a farmer's (i.e., borrower's) perspective, a lender's perspective, and a combination of the farmer's and lender's perspectives. The DP models are of a hog facility investment which costs $80,000. With this hog facility, 2000 hogs can be finished in a year. This investment is financed with either $20,000 of debt or $40,000 of debt. In either case, the loan is ten years in length, with loan obligations due at the end of each year.

The DP models evaluate the ability of the hog facility investment to be self liquidating. As such, funds from other enterprises are not considered. This focus ignores possible flows from other farm enterprises which may provide funds to repay loan obligations. However, since the flexibilities of loan plans are being evaluated, this perspective is judged as being appropriate.

Four DP models containing alternative loan repayment plans are solved. The first model, which is entitled "BASIC DP", models the firm with no debt financing. Results are used as a benchmark for the remaining loan repayment plans. The remaining three DP models are modifications of the BASIC DP model and incorporate the traditional amortization, flexible amortization, and variable amortization loan repayment plans. Each of the four models are described in the following sub-sections.

The Basic Dynamic Programming Model

Stages of the BASIC DP model are monthly periods. Twelve monthly periods are grouped in order to account for yearly income tax obligations. The BASIC DP model contains three state variables: per hog direct returns, taxable income, and debt/saving balance. Per hog direct returns is stochastic while the remaining state variables are deterministic. Per hog direct returns equals gross revenue from the sale of a 220 pound hog less variable costs of raising a hog from 40 pounds to 220 pounds. Taxable income accumulates income during a year. The debt/saving balance incorporates liquidity and additional debt requirements into the model. A positive balance results when funds generated from the hog operation have not been withdrawn. Positive balances can be used to counter cash shortfalls in future years. Negative amounts indicate that additional debt has been required to cover adverse incomes resulting from low or negative hog returns. The decision variable is the amount of funds to withdraw at the end of the year. This withdrawal can be used to either increase the size of the farming operation, make off-farm investments, or cover family living expenses.

The BASIC DP model can be written as follows:

(1-a) $V_t(HR_t, TI_t, B_t) = \max_{W_t} E\{R_t(HR_t, TI_t, B_t) + \beta \cdot V_{t+1}(HR_{t+1}, TI_{t+1}, B_{t+1})\}$

   subject to:

(1-b) $HR_{t+1} = f_1(HR_t)$

(1-c) $TI_{t+1} = f_2(TI_t, HR_t)$

(1-d) $B_{t+1} = f_3(B_t, TI_t, HR_t)$

where $V_t(\cdot)$ is the recursive objective function, HR_t is the per hog direct return state variable, TI_t is the taxable income state variable, B_t is the debt/saving balance state variable, $E\{\cdot\}$ is an expectations operator, $R_t(\cdot)$ is the current returns function, β is the discount factor, W_t is the amount of withdrawals, $f_1(\cdot)$ is the stochastic state transition equation for per hog direct returns, $f_2(\cdot)$ is the taxable income state transition equation, and $f_3(\cdot)$ is the debt/saving balance state transition equation.

The Recursive Objective Function. The current returns function within the recursive objective function given in (1-a) equals:

(2) $R_t(HR_t, TI_t, B_t) = \lambda \cdot W_t - (1 - \lambda) \cdot OUT_t$

where λ is a parameter which ranges between 0 and 1 and OUT$_t$ equals the amount of additional debt required at the end of the year. Specifically OUT$_t$ equals:

(3-a) $B_{t+1} + B_t$ when B_{t+1} and B_t are less than zero,

(3-b) OUT$_t$ = $B_{t+1} + 0$ when B_{t+1} is less than zero and B_t is greater than or equal to zero,

(3-c) 0 when B_{t+1} is greater than or equal to zero.

Equation (3-a) gives additional debt capital requirements when additional debt capital has been needed in previous years, as indicated by a negative debt/saving balance (i.e., B_t is less than zero). Equation (3-b) gives additional debt capital requirements when debt capital has not been accumulated previously, as indicated by a positive B_t. Equation (3-c) indicates that additional debt is not needed when the debt/saving balance is positive.

The parameter λ in the current returns function (equation 2) captures differences in the objectives of farmers (i.e., borrowers) and lenders. A λ value of 1 indicates that all weight is placed on maximizing the present value of withdrawals. This is the farmer's perspective presuming risk neutrality. If, on the other hand, λ equals 0 all weight is placed on not having to obtain additional debt, the lender's perspective. A λ between 0 and 1 provides alternative weightings between the farmer's and lender's perspective.

Withdrawals at the end of the year, W_t, can not occur if the resulting debt/saving balance, B_{t+1}, is negative. This restriction prevents withdrawals from the asset base by using debt capital. The discount rate equals .13 percent, the average percent return from the hog investment facility.

Hog Returns State Transition Equation. The stochastic, per hog direct return state transition is estimated using monthly data from the *Livestock and Meat Situation and Outlook* (USDA). These returns are adjusted to reflect mid-west conditions.

Evaluation of various time-series models suggest that an AR(2) model adequately models the series' time dependent nature. Resulting parameter estimates and t-statistics (in parentheses) are:

(4-a) $HR_{t+1} = 2.430 + 1.177 \cdot HR_t - .4393 \cdot HR_{t-1}$
       (3.60)   (16.54)   (-6.29)

This equation has an adjusted R-square of .7360 and a standard error of estimate of 7.239. Residuals show no sign of auto-correlation and the hypothesis of non-normally distributed residuals is rejected using the Jarque-Bera test statistic.

To reduce the dimensions of the DP models, only one direct return variable is included in the model. Burt and Taylor's method of reducing an auto-regressive process results in:

(4-b) $HR_{t+1} = 1.6888 + .8177 \cdot HR_t$

The reduced form has a standard error estimate of 6.6279.

Taxable Income State Transition Equation.   The deterministic, taxable income state transition equation accumulates taxable income during a year. This transition equation can be written as:

(5-a)

$$TI_{t+1} = \begin{cases} 0 \text{ for month twelve} \\ \\ TI_t + HR_t \cdot 167 - FC/12 + I_1(TI_t) \cdot TI_t \quad \text{otherwise} \end{cases}$$

(5-b)

where FC equals fixed costs, and $i_1(\cdot)$ is a function giving an interest rate. Equation (5-a) indicates that taxable income will equal zero at the beginning of each year. Ending year taxable income is withdrawn from the hog operation or flows into the debt/saving balance. Equation (5-b) gives taxable income changes between months during a year.

During a month, taxable income increases due to recognition of revenues and costs from hog sales, HR_t*167. The 2,000 hogs marketed during a year are presumed to move evenly through the hog facility. Thus, per hog direct returns are multiplied by 167, resulting in monthly revenues and variable costs. Fixed costs (FC) are presumed to occur evenly throughout the year. Thus, fixed costs are divided by 12 to arrive at monthly fixed costs. This amount, FC/12, reduces taxable income each month. Total fixed costs are adapted from *Ohio Livestock Enterprise Budgets, 1987* and equal $4,000 per year.

The final term of (5-b), $i_1(TI_t)*TI_t$, gives returns on financial holdings or costs on operating debt. Positive taxable incomes from previous months are presumed to be placed in a saving account yielding a 5 percent return. Thus, $i_1(\cdot)$ equals .05 when TI_t is positive. Negative taxable incomes are presumed to be covered by an operating loan. The interest rate on the operating loan equals 11 percent.

Debt/Saving Balance State Transition Equation.   The deterministic, debt/saving balance state transition equation given in (1-d) can be rewritten as:

(6-a)

$$B_{t+1} = \begin{cases} B_t \text{ for months one through eleven} \\ \\ B_t + TAX\{TI_t + HR_t \cdot 167 - FC/12 + i_1(TI_t) \cdot TI_t \\ + i_2(B_t) \cdot B_t\} - W_t \text{ for month twelve} \end{cases}$$

(6-b)

where TAX($\cdot$) is a function giving after-tax income and $i_2(\cdot)$ is a function giving the interest rate on the debt/saving balance. Equation (6-a) indicates that the debt/saving balance does not change during the year. Equation (6-b) gives the debt/saving balance change at the end of the year.

At the end of the year, the debt/saving balance equals the previous debt/saving balance plus taxable income during the year, TAX($\cdot$), less withdrawals, W_t. Terms within TAX($\cdot$) give before-tax income. Before tax income includes:

1) accumulated taxable income from the previous eleven months, TI_t,
2) taxable income during month twelve, $HR_t^*167 - FC/12 + i_1(TI_t)^*TI_t$,
3) returns or costs on the debt/saving balance, $i_2(B_t)^*B_t$. The function $i_2(\cdot)$ gives the yearly interest rate on the debt/saving balance. For positive balances, a 5 percent rate of return is used. Five percent equals the average saving rate obtainable at a commercial bank. For negative balances, an 10.5 percent interest rate is used to determine interest costs. This percent equals the average interest rate on intermediate term debt over the last 10 years.

Taxes are determined based on the 1988 tax code. These taxes include two federal brackets, a 12 percent social security tax rate, and five Ohio tax brackets.

Note that the final term in (6-b) is withdrawals, W_t. This is the decision variable and reduces the debt/saving balance. The implicitly assumed return on withdrawals is 13 percent, the discount rate. In the case of positive balances, the decision variable's tradeoff is between a 13 percent and a 5 percent return on debt/saving balance. If withdrawals occur, however, these funds are not available to repay debt. On the other hand, positive debt/saving balance can be used to repay debt. Thus, the debt/saving balance represents a liquid asset while withdrawals represents a non-liquid asset.

The Amortization Dynamic Programming Model
(AMOR DP) Model

The AMOR DP model includes a traditional amortization loan which is used to finance the acquisition of the hog facility. The loan is ten years in length and has constant, yearly principal and interest payments. Two loan sizes are used: $20,000 and $40,000. Yearly principal (PRIN$_t$) and interest payment (INT_t) for the two loan sizes are shown respectively in Panels A and B of Table 11.1.

Inclusion of the amortization loan does not require additional state variables. The debt/saving balance state transition equation, however, has to be modified to account for the principal and interest payments. Specifically, the debt/saving balance state transition equation is:

$$
\begin{aligned}
(7\text{-}a) \quad & \\
B_{t+1} = & \begin{cases} B_t \quad \text{for months one through eleven} \\[1em] B_t + TAX\{TI_t + HR_t \bullet 167 - FC/12 + i_1(TI_t) \bullet TI_t \\ + i_2(B_t) \bullet B_t - INT_t\} - PRIN_t - W_t \quad \text{for month twelve} \end{cases} \\
(7\text{-}b) \quad &
\end{aligned}
$$

where INT_t equals the yearly interest payment and $PRIN_t$ equals the yearly principal payment. The loan is presumed to be paid during the first ten years. During these years, the debt/saving state transition equation varies due to the differing principal and interest payments. Thus, the decision rule does not converge during the first ten years.

The Flexible Amortization Dynamic Programming (FLEX AMOR DP) Model

The FLEX AMOR DP model modifies the terms of the previously described amortization loan. Under the flexible amortization loan terms, principal payments can be paid ahead of schedule.

The specification of the FLEX AMOR DP model requires an additional state variable giving the principal outstanding on the amortization loan (P_t). Also, an additional decision variable is required. This decision variable is the amount of principal paid each year (PP_t). The recursive equation for the model then becomes:

$$
(8) \quad V_t(HR_t, TI_t, B_t, P_t) = \max E\{R_t(HR_t, TI_t, B_t, P_t) \\
+ \beta \bullet V_{t+1}, TI_{t+1}, B_{t+1}, P_{t+1})\} W_t, PP_t
$$

The decision variables are restricted such that unscheduled principal payments and withdrawals do not result in a negative debt/saving balance.

The debt/saving balance state transition equation also has to be modified:

$$
\begin{aligned}
(9\text{-}a) \quad & \\
B_{t+1} = & \begin{cases} B_t \quad \text{for months one through eleven} \\[1em] B_t + TAX\{TI_t + HR_t \bullet 167 - FC/12 + i_1(TI_t) \bullet TI_t \\ + i_2(B_t) \bullet B_t - .105 \bullet P_t\} - PP_t - W_t \quad \text{for month twelve} \end{cases} \\
(9\text{-}b) \quad &
\end{aligned}
$$

Table 11.1   Principal Outstanding, Interest Payment, and Principal Payment for Two Amortization, Ten Year Loans

Panel A. Beginning Debt = $20,000

Year	Principal Outstanding	Interest Payment	Principal Payment
0	20,000		
1	18,775	2,100	1,225
2	17,421	1,971	1,354
3	15,925	1,829	1,496
4	14,272	1,672	1,653
5	12,446	1,499	1,827
6	10,427	1,307	2,018
7	8,197	1,095	2,230
8	5,732	861	2,464
9	3,009	602	2,723
10	0	316	3,009

Panel B. Beginning Balance = $40,000

Year	Principal Outstanding	Interest Payment	Principal Payment
0	40,000		
1	37,550	4,200	2,450
2	34,842	3,943	2,708
3	31,850	3,658	2,992
4	28,544	3,344	3,306
5	24,891	2,997	3,653
6	20,854	2,614	4,037
7	16,394	2,190	4,461
8	11,465	1,721	4,929
9	6,018	1,204	5,446
10	0	632	6,018

Furthermore, an additional state transition equation has to be added to the model giving principal balance changes on the amortization loan. This state transition equation equals:

(10) $P_{t+1} = P_t - PP_t$

Note that the principal outstanding depends on the PP_t decision variable.

The Variable Amortization Dynamic Programming (VAR AMOR DP) Model

The VAR AMOR DP model incorporates a debt reserve in the amortization loan. The debt reserve is presumed to equal 10 percent of the loan outstanding. This addition does not require modification of the FLEX AMOR DP model's state transition equations. The size of the amortization loan and the range on the debt repayment variable, PP_t, does have to be modified. Under the flexible amortization loan, the largest amortization principal balance is $20,000 and $40,000, respectively, for the $20,000 and $40,000 amortization loans. Under the variable amortization loan, these sizes are increased by 10 percent to account for the debt reserve. The range on the decision variables allow use of the debt reserve. Thus, at the end of year one, for example, it is possible to borrow an additional 10 percent of the outstanding balance to cover adverse income outcomes.

Solving the Dynamic Programming Models

Numeric solutions are generated, requiring all state variables to be discretized. Per hog direct returns have 6 states ranging in equal increments from -$20 to $40. Taxable income ranges from -$60,000 to $60,000 in $15,000 increments, resulting in 9 states. The debt/saving balance ranges from -$400,000 to $400,000 in $10,000 increments, resulting in 81 states. This yields a total of 4,374 states for the BASIC and AMOR DP models. In addition to the above states, the FLEX AMOR and VAR AMOR DP models have 11 additional states associated with the principal outstanding state variable. These states are divided such that each represents a principal payment under the amortization plan. The second column of panels A and B respectively show the states for the $20,000 and $40,000 loan for the FLEX AMOR DP model. These additional states result in a total of 48,114 states for the FLEX AMOR and VAR DEBT DP models.

All models are recursively solved beginning at the final period. Linear interpolation of the objective function is used for the taxable income and debt/saving balance state variables. This is done in order to reduce biases resulting from discrete states and to increase the convergence rate. The BASIC DP program is recursively solved for five years, at which point the optimal decision rules converge. During the period in which the loan existed, the remaining models do not converge because the debt/saving balance state transition equation varies over the years of the amortization loans. To account for non-convergence, the remaining models are solved for a fifteen year period. During the first five years--

the final five years of the time frame--decision rules are generated. This results in a converged decision rule similar to that from the BASIC DP model. Then, decision rules for the ten year loan repayment period are generated.

When evaluating the alternative loan repayment plans, state variable distributions resulting from following the optimal decision rule are more useful than the decision rules. Therefore, optimal decision rules and the state transition equations are used to construct future state variable probability distributions following conditional probability methods (see Howard for a discussion). By using these methods, discrete joint probability density functions of the per hog direct returns, taxable income, and debt/saving balance can be ex ante forecasted, conditional on initial state variable levels and presuming that the optimal decision rules are followed. From the joint probability density function, discrete marginal distributions of a single state variable can be found. These marginal distributions then can be used to calculate expected values (see Schnitkey, or Novak and Schnitkey for a more detailed discussion of these methods).

Yearly conditional probability distributions have been computed using the same state variable distribution as in the dynamic programming models. Unless noted otherwise, all conditional probabilities are calculated using initial state intervals of a $10 per hog direct returns, a $0 taxable income, and a $0 debt/saving balance. The alternative loan repayment plans are evaluated using the following criteria: (1) the yearly expected values of withdrawals, (2) the yearly expected values of the debt/saving balance, (3) the probabilities of having additional debt (i.e., having a negative debt/saving balances), and (4) the marginal debt/saving balance distributions at the end of the amortization loans (in year 10).

Results

Results from the dynamic programming analysis are reported in the following four sub-sections. The first sub-section presents results for the BASIC DP model for λ values of 1, .33, and 0. The remaining three sub-sections then respectively give results for λ values of 1, .33, and 0 across the differing debt levels and loan repayment plans.

Results from the BASIC DP Model

Results from the BASIC DP model are of interest for two reasons. First, they provide a basic understanding of the dynamic factors at work.

Second, given similar λ values, each of the remaining models' decision rules are the same as the BASIC DP models' decision rules after the ten year amortization loans have expired.

Converged optimal decision rules from the BASIC DP models can be described based on the debt/saving balance that is maintained. These balances vary depending on the level of λ. For λ = 1, withdrawals occur whenever positive debt/saving balances exist. These withdrawals equal the amount of the positive debt/saving balances. For λ = .33, withdrawals occur such that a $30,000 saving balance is built and maintained. Any funds which would result in debt/saving balances above $30,000 are withdrawn. For λ = 0, a positive debt/saving balance of $350,000 is built. Once debt/saving balances reach this level, withdrawals occur.

Expected yearly withdrawals and debt/savings balances respectively are shown in panels A and B of Figure 11.1 for the alternative λ values. (All figures have been placed at the end of the chapter.) In early years, withdrawals are higher for higher λ values. This occurs because smaller savings balances are built to counter adverse outcomes. In later years, however, expected withdrawals are higher for lower λ values. This occurs because lower λ values have built saving balances that generate 5 percent returns. This produces greater taxable income, allowing larger withdrawals.

Note that consumption withdrawals trend downward from year 2 onward for λ = 1 and from year 5 onward for λ = .33 value. This occurs because an increasing probability is in the debt region of the debt/saving balance. Transition matrices resulting from the BASIC DP model are not ergodic (see Howard for a discussion). A trapping state exists at approximately a -$350,000 debt/saving balance. At this debt level, interest costs exceed taxable income for any combination of per hog direct returns. Thus, more debt is continually accumulated. This trapping state is equivalent to bankruptcy. This explains the downward trends in withdrawals (panel A of Figure 11.1) and debt/saving balance (panel B) for the two models. The rate at which withdrawals and debt levels decline depends on the initial debt level. This is illustrated in Figure 11.2 which shows expected withdrawals and debt/saving levels for λ = .33 and beginning debt/saving balances of $0 and $40,000.

The BASIC DP model with λ = 0 has two trapping states: one at the -$350,000 debt/saving balance (i.e., bankruptcy) and the other at a $350,000 debt/saving balance. Once a $350,000 debt/saving balance has been reached, no probability exists of falling into debt. Thus, the firm never goes bankrupt. For beginning state intervals of a $10 per hog direct return, a $0 taxable income, and a $0 debt/saving balance, the convergent probability of bankruptcy equals approximately 3 percent.

Results for a 1.0 λ Value

Since the decision rules from each model vary over the years of the loan, the optimal decision rules are not presented. Instead conditional probabilities for beginning state levels of a $10 per hog direct return, a $0 taxable income, and a $0 debt/saving balance are shown. This allows for a basic understanding of the optimal decision rules. In addition, conditional probabilities allow the performance of each loan repayment plan to be analyzed.

Figure 11.3 shows the yearly expected withdrawals from the AMOR and VAR AMOR DP models. Results from the FLEX AMOR DP model are not presented because they are the same as the AMOR model results. This indicates λ levels of 1.0 do not result in any prepayment of principal. Panel A of Figure 11.3 shows expected withdrawals for a $20,000 beginning amortization debt level while panel B gives expected withdrawals for a $40,000 beginning level.

In all years up to year 10, the VAR AMOR model results in higher expected withdrawals. This is primarily due to the flexibility of the variable amortization loan plan. At the same time, expected debt/saving balances indicate that lower debt results from the variable amortization loan (Figure 11.4). In year 7, for example, approximately $20,000 of debt is needed for the amortization loan given a beginning debt level of $20,000. Additional debt of $14,000 is needed for the variable amortization loan.

Higher debt levels under the amortization loan are primarily due to a higher probability of having additional debt, as shown in Figure 11.5. The yearly probabilities in Figure 11.5 give the probabilities of having negative debt/saving balances. Negative debt/saving balances result from cash shortfalls. The terms of variable amortization loan allow for more of the cash shortfalls to be covered by the amortization loan, thus the lower probabilities.

While the variable amortization loan results in lower probability of having debt in years 1 through 9, it does not result in lower probabilities in year 10, the final year of the amortization loan. This is illustrated in Table 11.2 which contains two panels giving results for the $20,000 amortization loan and the $40,000 amortization loan. For each loan size and the three loan repayment plans, the expected ending debt/saving balance, expected future value of withdrawals, and expected debt/saving balance plus withdrawals is given. Expected future value of withdrawals is the withdrawals from years 1 through 10 compounded to year 10. The expected debt/saving plus withdrawals can be used to evaluate the profitability of the loan repayment plans to the borrower. In addition, the debt/saving distribution in year 10 and the probability of having debt is given in each panel.

The variable amortization loan results in significantly higher expected debt/saving balances plus withdrawals than the other two loans. As stated before, however, the probability of having additional debt is higher under the variable amortization loan. For the $40,000 beginning amortization loan the probability of having debt is .1266 higher. At the same time, the variable amortization loan results in lower probability for debt/saving balances less than -$105,000. Approximately a .02 probability difference exists for both loan sizes.

Results for a .33 λ Value

Yearly expected withdrawals from the AMOR, FLEX AMOR, and VAR AMOR models are shown in Figure 11.6. Similar to results from the DP models having $\lambda = 1$, the variable amortization loan plan results in higher withdrawals in all years except in year 10. The variable amortization loan plan's withdrawals are above the flexible amortization loan plan until year 7.

Debt/saving balances are lower under the variable amortization loan plan than the other two loan plans until at least year 7 (Figure 11.7). This primarily results from the ability to retire debt instead of holding savings. Thus, the debt/saving distribution is partially truncated in the saving range. Note that the expected debt/saving balance is distinctly different between the flexible amortization and variable amortization loan plans, even though the flexible amortization plan allows the amortization loan's principal to be prepaid. Differences result from the debt reserve under the variable amortization plan, which allows cash shortfalls to be covered.

As a result, the probability of having additional debt is lower under the variable amortization loan plan (Figure 11.8). Note also the probability of having additional debt differs little between the amortization and flexible amortization loan plans. This suggests that the debt reserve under the variable amortization loan plan is key in providing flexibility in meeting cash shortfalls.

Table 11.3 presents summary conditional probability results in year 10. Similar to results when $\lambda = 1$, the expected debt/saving balance plus withdrawals is higher for the variable amortization plan than it is for the other two plans. Also, the probability of having a debt/saving balance less than -$55,000 is lower under the variable amortization plan. Unlike the case when $\lambda = 1$, however, the probability of being in debt is lower.

Table 11.2 Expected Debt, Future Value of Withdrawals and Debt/Saving Distribution in Year 10 for Differing Debt Instruments and Beginning Debt Levels, Lambda = 1.0

	Debt Instrument[a]		
	Amor.	Flexible Amor.	Variable Amor.
Panel A. Beginning Debt = $20,000			
Expected Ending Debt(-)/Saving (+)	-28,831	-28,831	-27,803
Expected Future Value of Withdrawals	168,248	168,248	186,055
Expected Debt/Saving plus Withdrawals	140,000	140,000	158,252
Debt/Saving Distrib. (yr. 10)	----Probability-----		
less than -$105,000	.0740	.0740	.0581
-$105,000 to -$55,000	.1219	.1219	.1242
-$55,000 to -$5,000	.3318	.3318	.4099
-$5,000 to +$5,000	.1236	.1236	.3621
$5,000 to $25,000	.0485	.0485	.0457
$25,000 to $55,000	.0000	.0000	.0000
Probability of Having Debt	.5277	.5277	.5922
Panel B. Beginning Balance = $40,000			
Expected Ending Debt(-)/Saving (+)	-44,361	-44,361	-45,755
Expected Future Value of Withdrawals	131,091	131,091	163,113
Expected Debt/Saving plus Withdrawals	86,730	86,730	117,369
Debt/Saving Distrib. (yr. 10)	----Probability-----		
less than -$105,000	.1371	.1371	.1182
-$105,000 to -$55,000	.1679	.1679	.1968
-$55,000 to -$5,000	.3578	.3578	.4744
-$5,000 to +$5,000	.3063	.3063	.1959
$5,000 to $25,000	.0309	.0309	.0147
$25,000 to $55,000	.0000	.0000	.0000
Probability of Having Debt	.6628	.6628	.7894

[a] See text for definition of differing debt instruments.

Table 11.3 Expected Debt, Future Value of Withdrawals and Debt/Saving Distribution in Year 10 for Differing Debt Instruments and Beginning Debt Levels, Lambda = .33

	Debt Instrument[a]		
	Ammor.	Flexible Ammor.	Variable Ammor.
Panel A. Beginning Debt = $20,000			
Expected Ending Debt(-)/Saving (+)	-3,364	-3,601	-1,897
Expected Future Value of Withdrawals	120,200	120,070	138,485
Expected Debt/Saving plus Withdrawals	116,836	116,469	136,588
Debt/Saving Distrib. (yr. 10)	----Probability------		
less than -$105,000	.0448	.0434	.0303
-$105,000 to -$55,000	.0854	.0812	.0766
-$55,000 to -$5,000	.1456	.1477	.1596
-$5,000 to +$5,000	.0747	.0813	.0949
$5,000 to $25,000	.4875	.5225	.5432
$25,000 to $55,000	.1610	.1232	.0950
Probability of Having Debt	.2758	.2723	.2665
Panel B. Beginning Balance = $40,000			
Expected Ending Debt(-)/Saving (+)	-17,327	-16,373	-11,215
Expected Future Value of Withdrawals	80,040	80,950	97,390
Expected Debt/Saving plus Withdrawals	62,713	64,577	86,175
Debt/Saving Distrib. (yr. 10)	----Probability------		
less than -$105,000	.0881	.0868	.0599
-$105,000 to -$55,000	.0947	.0913	.0882
-$55,000 to -$5,000	.2008	.1918	.2048
-$5,000 to +$5,000	.0789	.0774	.0928
$5,000 to $25,000	.4399	.4550	.4823
$25,000 to $55,000	.0976	.0997	.0720
Probability of Having Debt	.3836	.3699	.3529

[a] See text for definition of differing debt instruments.

Results for a 0.0 λ Value

Yearly expected withdrawals, debt/saving balances, and probability of having additional debt are similar between the three loan repayment plans at a 0 λ value are similar to those shown for the λ = .33. Thus, they are not shown. Only results from year 10 are shown in Table 11.4.

Table 11.4 **Ending Debt, Future Value of Withdrawals and Debt/Saving Distribution in Year 10 for Differing Debt Instruments and Beginning Debt Levels, Lambda = 0.0**

	Debt Instrument[a]		
	Ammor.	Flexible Ammor.	Variable Ammor.
Panel A. Beginning Debt = $20,000			
Expected Ending Debt(-)/Saving (+)	77,381	78,654	89,315
Expected Future Value of Withdrawals	3,548	3,531	3,876
Expected Debt/Saving plus Withdrawals	80,929	82,185	93,191
Debt/Saving Distrib. (yr. 10)	----Probability-----		
less than -$105,000	.0397	.0394	.0257
-$105,000 to -$55,000	.0463	.0459	.0357
-$55,000 to -$5,000	.0863	.0846	.0722
-$5,000 to +$5,000	.0246	.0248	.0213
$5,000 to $25,000	.0576	.0560	.0520
$25,000 to $55,000	.1074	.1055	.1014
greater than $55,000	.6381	.6438	.6917
Probability of Having Debt	.1723	.1699	.1336
Panel B. Beginning Balance = $40,000			
Expected Ending Debt(-)/Saving (+)	39,777	41,545	56734
Expected Future Value of Withdrawals	1,240	1,296	1,590
Expected Debt/Saving plus Withdrawals	41,017	42,841	58,324
Debt/Saving Distrib. (yr. 10)	----Probability-----		
less than -$105,000	.0838	.0833	.0556
-$105,000 to -$55,000	.0797	.0788	.0638
-$55,000 to -$5,000	.1281	.1252	.1120
-$5,000 to +$5,000	.0327	.0323	.0300
$5,000 to $25,000	.0728	.0719	.0689
$25,000 to $55,000	.1231	.0997	.1213
greater than $55,000	.4798	.4877	.5484
Probability of Having Debt	.2916	.2873	.2314

[a] See text for definition of differing debt instruments.

These results again are similar to those shown for $\lambda = .33$: the expected debt/saving balance plus withdrawals is higher under the variable amortization plan, the probability of having debt is lower under the variable amortization plan, and the probability of having debt/saving balances below -$55,000 is lower under the variable amortization plan.

Summary of Results

Results from the DP models can be summarized as follows:

1. For all λ values and initial amortization debt levels, the variable amortization loan results in higher expected withdrawals plus ending debt/saving balances than the other two repayment plans. Thus, the variable amortization loan is more profitable from a borrowers standpoint, regardless of the level of emphasis given to the lender's objectives.
2. For λ values of .33 and 0, the probability of having debt is lower and the expected debt/saving balance is higher under the variable amortization loan plan than under the other two loan plans. This suggests that the variable amortization loan plan has advantages for lenders.
3. At a λ value of 1, the variable amortization loan plan results in higher probabilities of having debt and smaller debt/saving balances than the other two loan plans. This suggests that lenders would not prefer this loan plan for profit maximizing, risk neutral borrowers.
4. Loan size does not influence the direction of results. Thus, debt levels do not seem to be a factor in preferring one loan plan over another.

Conclusions

Stochastic dynamic programming models have been solved which analysis the performance of three loan repayment plans: a traditional amortization loan, a flexible amortization loan, and a variable amortization loan. Performance has been monitored using differing loan sizes and objectives representing a borrower's perspective, a lender's perspective, and a combination of a borrower's and lender's perspective. Results indicate that borrowers may prefer the variable amortization loan because it has a debt reserve. This reserve serves as a liquidity source during periods of adverse income. Results also indicate that lenders may prefer the variable amortization plan, given that some restrictions are placed on withdrawals of the borrower.

These results suggest that the variable amortization has potential as a viable loan instrument in the agriculture sector. From the lenders perspective, loan terms have to be scheduled such that additional debt is not generated by profit-maximizing farmers. This is an area requiring further research. Moreover, terms of the loan have to be simplified such that they are easily implementable.

Notes

1. An objective of minimizing the probability of cash shortfalls also is appropriate. The dynamic programming models were solved using this objective. Similar results were obtained.

References

Baker, C.B. "A Variable Amortization Plan to Manage Farm Mortgage Risk." *Agricultural Finance Review* 36(1976):1-6.

Baker, C.B. *Structural Issues in U.S. Agriculture and Farm Debt Perspectives.* The University of Kansas Law Review. 34(1986):457-67.

Burt O.R., and C.R. Taylor. *Reduction of State Variable Dimension in Stochastic Dynamic Optimization Models Which Use Time Series Data.* University of California, Davis, Department of Agricultural Economics, Working Paper No. 88-2, 1988.

Howard, R. *Dynamic Programming and Markov Processes.* New York: John Wiley and Sons, 1960.

Lee, W.F. and C.B. Baker. "Agricultural Risks and Lender Behavior." in *Risk Management in Agriculture.* P.J. Barry, editor. Iowa State University Press, Ames, Iowa. 1984.

Lee, W.F. "Some Alternatives to Conventional Farm Mortgage Loan Repayment Plans." *Canadian Journal of Farm Economics* 14(1979):12-20.

Novak, F.S., and G.D. Schnitkey. "Dynamic Investment Models: An Application to Agriculture." Paper presented at the "Applications of Dynamic Programming to Agricultural Decision Problems" conference, Auburn University, Auburn, Alabama, May 9-10, 1989.

Schnitkey, G.D. "A Dynamic Programming Analysis of Optimal Farmland Investment Decisions: An Application to Central Illinois High-Quality Farmland. Ph.D. thesis, University of Illinois, 1987.

U.S. Department of Agriculture. *Livestock and Meat Situation and Outlook.* E.R.S., selected issues.

Panel A. Expected Withdrawal Per Year

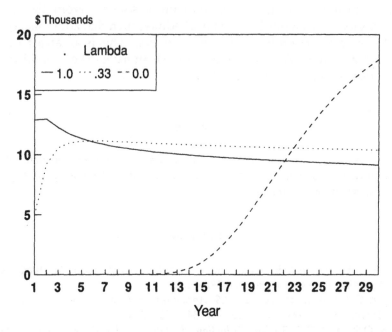

Panel B. Expected Debt/Saving Balance
Per Year

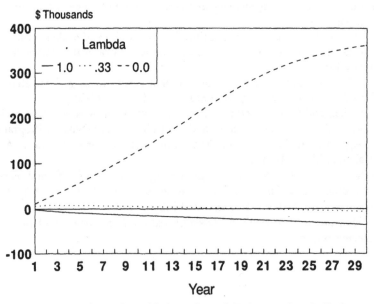

Figure 11.1 Expected Yearly Withdrawals and Debt Levels, No Debt

Panel A. Mean Withdrawal Per Year

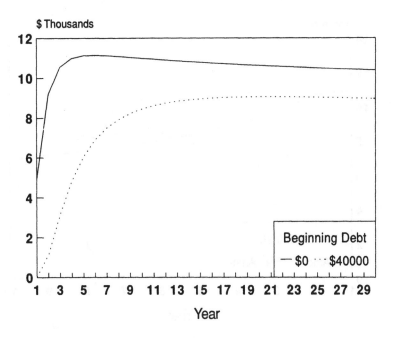

Panel B. Mean Debt/Saving Balance
Per Year

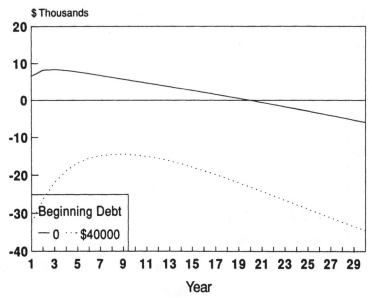

Figure 11.2 Expected Yearly Withdrawals and Debt Levels, Lambda = .33

Panel A. Beginning Debt = $20,000

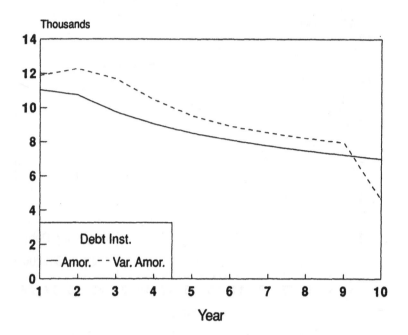

Panel B. Beginning Debt = $40,000

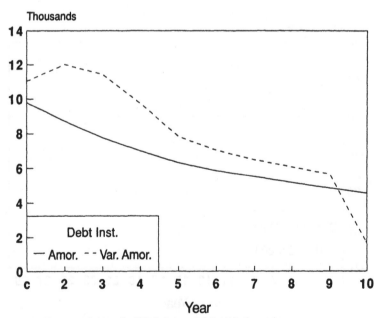

Figure 11.3 Expected Yearly Withdrawals, Lambda = 1.0

Panel A. Beginning Debt = $20,000

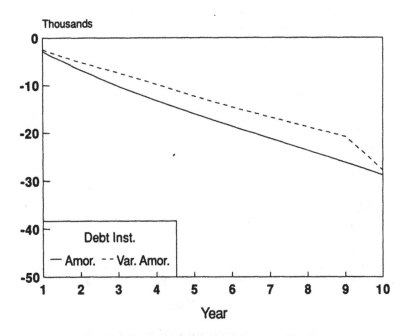

Panel B. Beginning Debt = $40,000

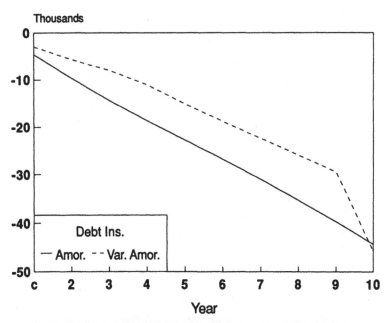

Figure 11.4 Expected Yearly Debt/Saving Balance, Lambda = 1.0

Panel A. Beginning Debt = $20,000

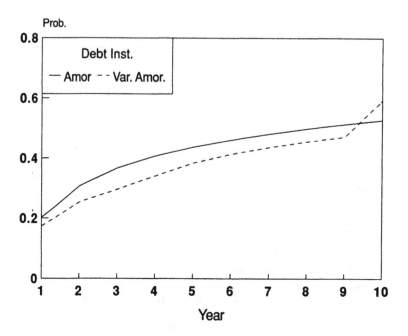

Panel B. Beginning Debt = $40,000

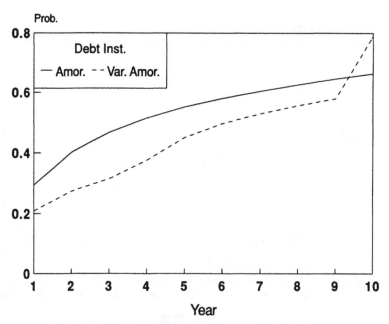

Figure 11.5 Probability of Having Additional Debt, Lambda = 1.0

Panel A. Beginning Debt = $20,000

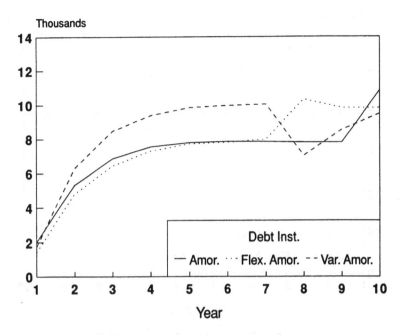

Panel B. Beginning Debt = $40,000

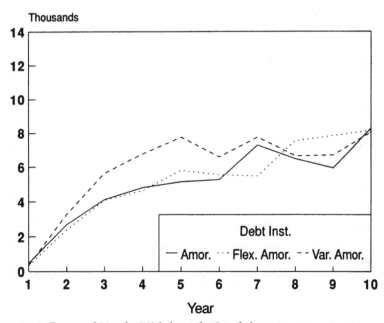

Figure 11.6 Expected Yearly Withdrawals, Lambda = .33

Panel A. Beginning Debt = $20,000

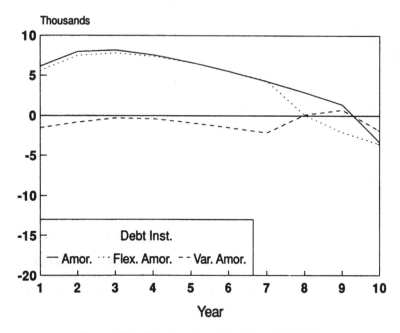

Panel B. Beginning Debt = $40,000

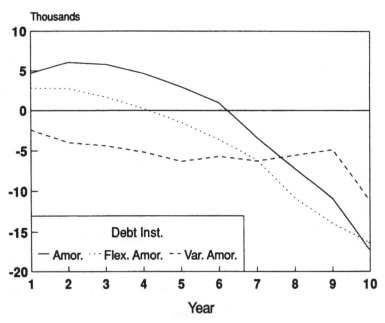

Figure 11.7 Expected Yearly Debt/Saving Balance, Lambda = .33

Panel A. Beginning Debt = $20,000

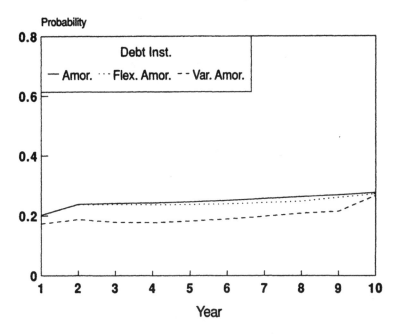

Panel B. Beginning Debt = $40,000

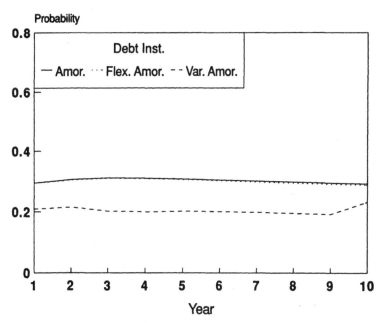

Figure 11.8 Probability of Having Additional Debt, Lambda = .33

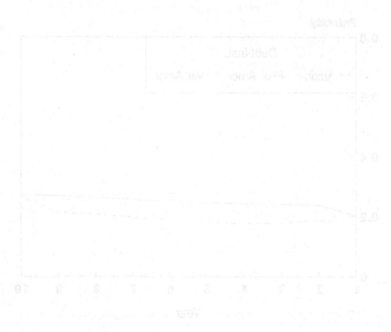

Panel A. Beginning Debt = $20,000

Panel B. Beginning Debt = $40,000

Figure 11.6 Probability of Having Additional Debt, Lambda = 33

T - #0042 - 071024 - C0 - 222/152/11 [13] - CB - 9780367011055 - Gloss Lamination